HIGGS FORCE

1777

Other titles by Nicholas Mee

Multimedia CD-ROMs

POLYTOPIA

Art and Mathematics

Life, the Universe and Mathematics

Symbolic Sculpture
John Robinson and Nicholas Mee

The Code Book on CD-ROM
Simon Singh and Nicholas Mee

Maths Lesson Starters
David Benjamin, Justin Dodd and Nicholas Mee

Nubble!
Edgar Fineberg, Jack Berkovi and Nicholas Mee

Connections in Space
Nicholas Mee, John Barrow, Martin Kemp and Richard Bright

Key Concepts in Chemistry
Debra Nightingale and Nicholas Mee

www.virtualimage.co.uk

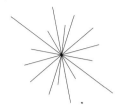

HIGGS FORCE

Cosmic Symmetry Shattered

The story of the greatest scientific
discovery for 50 years

Nicholas Mee

Quantum Wave Publishing
www.quantumwavepublishing.com

Published by

Quantum Wave Publishing Limited

Bowden Hall, Bowden Lane, Marple SK6 6ND,
United Kingdom

www.QuantumWavePublishing.com
enquiries@QuantumWavePublishing.com

ISBN (PB): 978-0-9572746-1-7
ISBN (HB): 978-0-9572746-2-4
E-ISBN: 978-0-9572746-0-0

British Library Cataloguing in Publication Data
A record is available from the British Library

First published 2012 by Lutterworth Press
Second edition published 2012 by Quantum Wave Publishing

Contents

INTRODUCTION

BANG!

In the first instant of creation, the universe was an incredibly hot and fantastically dense fireball. The universe was born with perfectly symmetrical crystalline laws, and a single force that determined the behaviour of its constituents. Within a moment the temperature began to fall as the universe expanded and the perfect symmetry was lost. The original force was shattered into several disparate pieces. One force would hold the quarks together in the atomic nucleus. One force would transmute matter and make the different elements. And one force would bind the atoms and control their chemical reactions.

The modern synthesis of particle physics can explain how all of this happened 13.7 billion years ago and how the same laws control the behaviour of matter today. Unfortunately these esoteric secrets are known to but a few initiates, the High Priests of Particle Physics. They are hidden behind a veil of abstract and subtle mathematics. But I will be drawing back the veil and revealing the secrets of the universe to anyone who cares to take a look.

Physicists have discerned a beautiful and symmetrical architecture beneath the apparent chaos of the world in which we live. The universe is constructed according to an elegant geometrical design based on a few simple principles. However, if the universe displayed this perfect symmetry in its entirety, it would be completely unchanging, formless and barren. This is not at all what our universe looks like. We live in a universe that is dynamic and full of activity. So here we have a profound enigma: if the structure of matter is organised in an orderly and symmetrical fashion at its deepest level, where does the diversity and complexity of the world come from? Physicists now think they have the answer.

In recent decades, physicists have pieced together an amazingly accurate and beautiful theory which binds together all the known elementary particles and the forces that act between them. This theory represents the culmination of philosophical and scientific attempts to understand the structure of matter dating back over two thousand years. The predictions of the theory, including the existence of new elementary particles, have been tested in accelerator experiments. So far the theory has passed every test without a flaw. Until now there was just a single piece missing from the jigsaw – a particle that is the key to the whole theory. It is this particle, known as the Higgs, that breaks the symmetry between the forces of nature and enables the universe to evolve into a complicated and interesting place. According to the theory, without the Higgs the universe would be empty and we would not be here to contemplate it. But is the theory correct? We now know the answer to this question. Physicists at the world's most powerful particle accelerator, the Large Hadron Collider (LHC), announced on 4 July 2012 that they have finally found the Higgs particle.[1]

It took ten years to construct the LHC, at a site straddling the Swiss/French border close to Geneva. This machine, which belongs to the European Organization for Nuclear Research (CERN), is the largest physics laboratory ever constructed. It is, in effect, the world's most powerful microscope and is now probing the constituents of matter in greater detail than has ever been achieved before. The collider accelerates two beams of protons to within a whisker of the speed of light before smashing them together and studying the resulting debris.[2] In order to surpass all previous particle accelerators and investigate new realms of physics, every feature of the LHC is record-breaking. It is an incredible technological achievement. Its discovery of the Higgs may be the first of many that will transform our understanding of the universe.

This book tells the story of the scientists who have elucidated the structure of matter, culminating in the amazing modern synthesis of the laws of nature into a single theory. It is the story of the fundamental constituents of matter themselves and the forces that bind them together. It is about the symmetry at the heart of matter, the mystery of how this symmetry is broken and the enormous challenge of completing the picture by tracking down the Higgs. Along the way we will meet a chemist who was addicted to the pleasures of laughing gas; the inventor of the kaleidoscope, whose business sense didn't match his scientific acumen; the ghostly apparition of the Brocken spectre, which inspired a device to reveal the paths of elementary particles; a physicist who compared his power to transmute the elements to that of the mythical alchemist Hermes Trismegistus; and an astronomer who was captivated by the beauty of a falling snowflake.

The first three chapters provide the broad historical and philosophical background. The next three describe, in turn, each of the forces that are important in particle physics. The final three chapters are about the modern synthesis of the particles and forces and the search for the last missing piece in the particle physics jigsaw.

Chapter 1 answers the question of why symmetry is important in describing the world around us. Whenever symmetries are revealed in the laws of nature, they provide a clue to a deeper order hidden within the structure of matter. This leads into Chapter 2, which is about the idea of unification: often a multitude of different phenomena have the same root cause. More specifically, at several times in the history of science physicists have found that what appear to be two or more different forces can best be described by a single unified force. We now know that just three different forces play a role in particle physics, and the goal of physicists is to reduce this number even further. Chapter 3 is about the crazy world of quantum mechanics, where particles behave like waves and waves behave like particles. It is all very strange, but this is what we find when we look at the structure of the atom. Quantum mechanics challenges our preconceptions of how the universe operates at its most intimate level.

The next three chapters describe each of the three fundamental forces. Chapter 4 is about the force that binds atoms together and controls chemical reactions. It is also the force that is powering my laptop – electromagnetism. Modern theories of particle physics are not just vague ideas: they agree with experimental measurements to an extraordinary precision. In this chapter we will see just how astonishing the match between theory and reality is. Chapter 5 is about the force that causes the transmutation of chemical elements within the stars. Without this force, known to physicists as the weak force, there would be no complex matter, no Earth and no people – the universe would consist of hydrogen atoms and nothing more. We are composed of the ash from the nuclear furnaces that roared at the hearts of previous generations of stars, and of the stardust that was dispersed throughout the heavens in past supernova explosions – ashes to ashes, dust to dust. In Chapter 6 we will see that although the universe looks very complicated and chaotic, it is constructed according to very simple laws. This chapter is about quarks and the force that binds them together into protons and neutrons, and ultimately holds the atomic nucleus together. This is the strong or colour force.

In the final part of the book all the strands are woven together into the tapestry of modern physics. Chapter 7 explains how, although physicists always seek the simplest and most symmetrical laws of nature, much of this symmetry must somehow be hidden for the universe to have evolved into such a complex and interesting place. The idea of symmetry breaking enables the electromagnetic and weak forces to be unified into a single electroweak force. This implies that there is a field – called the Higgs field – permeating the whole of space, through which all the particles in the universe are continually streaming. It breaks the symmetry between the forces and produces a new force – the Higgs force – that gives most of the elementary particles their mass. Chapter 8 shows how the modern synthesis of particle physics became established and reveals the conclusive experimental observations on which it is founded. We now have a succinct collection of all the elementary particles and the forces that bind them together. This close-knit family fits into a coherent scheme, rather like the periodic table of the chemical elements. With the discovery of that last missing particle – the Higgs – the table is now complete. The Higgs is crucial to the modern synthesis. It is the particle responsible for the symmetry breaking that took place in the first moments of the universe. Chapter 9 is about Europe's new particle accelerator, designed to probe the innermost secrets of matter and find the Higgs. Tracking down the Higgs has been one of the greatest achievements in the history of science. It confirms that we have an extremely good understanding of the functioning of the universe at its most fundamental level. A new chapter has now been opened, as physicists seek to answer even more ambitious questions about the structure of the universe.

Chapter One

SEEING THE WORLD THROUGH
KALEIDOSCOPE EYES

Picture yourself in a boat on a river,
With tangerine trees and marmalade skies,
Somebody calls you, you answer quite slowly,
The girl with kaleidoscope eyes.

John Lennon and Paul McCartney,
'Lucy in the Sky with Diamonds' (1967)

Cosmic Symmetry

One hundred metres beneath the ground on the outskirts of Geneva, two protons slam into each other in a mighty collision. There is nothing special about these two protons. They are indistinguishable from the untold myriads of these basic building blocks which form the atoms within our bodies and compose the world around us. Their only distinction is that they have just undergone an enormous acceleration inside the Large Hadron Collider, whirling around the giant machine and reaching energies beyond those ever before attained on Earth. For an instant, this collision recreates the conditions of the universe immediately after the Big Bang. In a tiny region of space, the forces of nature are reunited and some of the hidden symmetry in their laws is revealed. The tremendous release of energy produces a new particle, the Higgs, unseen since the very earliest moments of the universe.

As you read this, the Higgs has been making its first dramatic appearances in the detectors of the LHC. The early tantalising hints of its presence have been confirmed in the most spectacular fashion. Until now, it has remained encapsulated in the algebraic hieroglyphs of the world's leading particle physicists, one of whom – the British theorist Peter Higgs – gave his name to the particle. The purpose of the LHC is to breathe fire into his equations and create these first Higgs particles to exist on Earth. The Higgs is a key component of the theories that reign supreme today. It is a particle like no other, with a role that is fundamental to the creation of order in our universe. Although the universe began in a perfectly symmetrical state, the Higgs broke this symmetry and enabled the matter that formed within the universe to evolve into complex and diverse structures. It is as though the universe froze at a temperature of around a trillion degrees, with the result that its properties changed completely. Without the Higgs this transformation could not have happened, and the universe would have remained in a state that was homogeneous, lifeless and uninteresting. But, what do physicists mean when they say that a symmetry is 'broken'? And for that matter, what do they mean by 'symmetry'?

To fully appreciate the significance of the Higgs and understand how it fits into the beautiful and phenomenally accurate modern theories of the universe, first we need to shake off our preconceptions, return to a state of innocence and view the world as it would have appeared to pre-technological eyes. Throughout the ages, people have gazed up at the heavens and asked deep questions about their own existence and the structure of the world around them. We live in a fortunate age; modern physics can now provide precise answers to many of these fundamental questions. The answers are often encrypted in mathematical ciphers and unfamiliar terminology, but although some of the concepts of modern physics may be strange and even shocking, it is possible to express them in ordinary language, as I intend to demonstrate. The knowledge encoded in modern theories has been built up over many centuries. Since the 1600s there has been a huge acceleration in the development of our understanding of the universe, triggered by the establishment of experimentation as a means of rejecting theories that do not fit the facts. Each generation has built on the work of its predecessors, and many critical ideas can be traced all the way back to the philosophers of the ancient world.

Take a look at the world around you. The immediate impression is of variety, with objects of all colours, shapes and sizes. You will probably see a great deal of diversity, but little sign of pattern or regularity. It is in no way obvious that at their most fundamental level all the objects you can see have a similar composition, or that their innermost structure can be described in terms of symmetry and geometry. Most observers throughout history must have viewed the world in this way. Indeed, despite the technological superiority of China throughout most of human history, according to the great scholar of Chinese science, Joseph Needham,[1] Chinese philosophers did not develop an atomistic, geometrical or mechanical view of the material world, but saw it as organic. They considered natural processes, such as turbulent flowing streams or swirling veins in precious pieces of jade, to function like living entities. This world view was not conducive to the development of deterministic, rational laws of nature.

Some of the philosophers of ancient Greece saw things differently. They believed that beneath the surface of material objects lay patterns. From time to time they encountered symmetrical objects in nature: a pearl in an oyster shell, a flower in a water meadow, a starfish on the beach, a snowflake in a winter storm, a crystal in a rock face. Thinkers such as Pythagoras and Plato thought that these symmetrical objects were hinting at much deeper patterns and symmetries within matter, and they believed that numbers and geometry held the key to understanding them. With a great leap of imagination, they proposed that these relatively unusual but symmetrical natural forms might be offering a clue to the fundamental structure of all matter. They reasoned that the universe must be founded on elegant geometrical principles, and that by studying proportion, harmony and symmetry they could attain a glimpse of the divine order inherent in the universe's structure.

The Greek geometers had made a remarkable discovery: the concept of mathematical proof, which gave them a profound vision of their subject, radically different to that of their predecessors. Mathematics had thus been raised to a higher plane than all other subjects, and given an aura of elegance and the eternal. It may have been this new perspective on mathematics that prompted the philosophers to make their audacious proposals about the structure of the material world. For the ancient Greek mathematicians, their subject was a purely abstract enterprise that was worth pursuing only for the love of its intrinsic beauty. They were not interested in applying arithmetical methods to help keep their accounts or survey their property, and they were above dirtying their hands by grinding through tedious computations. What they sought was the sublime beauty of a timeless geometry that would offer insight into an abstract world constructed from pure reason, an orderly world that is ideal and uncorrupted. They believed there to be a body of eternal mathematical truths that could be deduced by following a trail of logical reasoning, starting with a few self-evident definitions or common notions that they called axioms. These were formal statements of what the geometers meant by terms such as 'line' or 'point', along with definitions of the most fundamental properties of these objects, such as what it means to say that two lines are parallel. Once the axioms were clearly enunciated, the rules of the game were established. From this point on everything else would follow by the application of reason alone.

The Greek knowledge of geometry was passed down the ages through Euclid's textbook, the elegant but austere *Elements*, whose influence continues today. The *Elements* presents the mathematics of classical Greece in a unified and rational style. It builds the whole edifice of geometry and mathematics upon a collection of self-evident axioms with a clarity and logic that has captivated philosophers and mathematicians ever since it was written in the third century BCE. Thomas Jefferson had the spirit of Euclid in mind when he drafted the American Declaration of Independence two thousand years later in 1776: 'We hold these truths to be self-evident, that all men are created equal, that they are endowed by their Creator with certain unalienable Rights, that among these are Life, Liberty and the pursuit of Happiness.'[2]

To the ancient Greeks, the power of mathematics was that it was a body of knowledge completely independent of the physical world. They reasoned about abstract ideal entities, not physical objects. For instance, when they discussed a circle they would define it logically as a collection of all the points in a plane that are situated at exactly the same specified distance from a central point. They were not concerned with whether it was possible to draw a perfect circle, or even whether any perfect circles actually exist. Although they might from time to time scratch a diagram of a circle in the sand to help them think about a problem, this representation of a circle was merely a useful schematic guide and not an attempt to draw a perfect circle. The mathematical truths that the Greek mathematicians pursued were truths about abstract mental images of perfect geometrical objects. The amazing thing about Greek mathematics is that once the rules of the game were set out, results could be proved beyond all doubt by pure reason. These results did not depend on accuracy of measurement, opinion, tradition – or even the weather. The proofs were eternal, and everyone who knew the rules of the game could agree on which results were true and which were false.

One simple example of a result that can be proved by logical deduction is the well-known fact that the angles of a triangle always add up to 180 degrees, or half a complete rotation.[3] This statement can be proved to follow logically from Euclid's axioms in just a couple of lines of reasoning. As a particular elementary fact about triangles, it may not be very interesting in itself, but what is striking is that it is possible to make such completely universal and eternal statements at all. This geometrical statement is true now and it will be true in a million years' time. It is true here and it is true in the Andromeda Galaxy. If there are any alien civilisations in the Andromeda Galaxy, then they will also know that the sum of the angles of a triangle is always equal to half a complete rotation. It is the absolute certainty of this statement that makes it so profound and so useful. Mathematicians know that this piece of information is rock-solid; they can see that it follows in a logical way from the self-evident axioms on which Euclidean geometry is built. They can then use it as a building block from which they can construct more complicated results. If the axioms are the foundation stones of the edifice of geometry, then this simple result might be likened to one of the floorboards of the ground floor.

Geometrical Alchemy

> Mathematics, rightly viewed, possesses not only truth, but supreme beauty – a beauty cold and austere, like that of sculpture.

Bertrand Russell, *Mysticism and Logic* (1919)[4]

Plato, who was born in the fifth century BCE, is still recognised as the greatest of all philosophers. In fact, the twentieth-century mathematician and philosopher Alfred North Whitehead claimed with some justification that the whole of Western philosophy was merely 'a series of footnotes to Plato'.[5] At the heart of Plato's philosophy lay the concept of mathematical proof. For him, it implied the existence of an abstract mental realm inhabited by perfect geometrical objects along with all the incontestable proofs about their properties and relationships. In

Plato's view, the physical world of the senses is a mere shadow of this transcendental and eternal geometrical world, which exists completely independently of reality. Furthermore, he believed that the architect of the universe had used the subtle geometry of this abstract universe as the blueprint when creating the physical universe. So, even if the real world appears to be mostly disorganised and unruly, at its core it is modelled on perfect abstract geometry and built upon elegant mathematical principles. Plato's conclusion was that symmetry and beautiful geometry must reside at the heart of matter. Most of the scientific notions of the ancient Greek philosophers appear hopelessly naive today. For instance, Anaximander is reputed to have believed that the sky was surrounded by a sphere of fire and that the Sun was simply a hole opening onto this sphere, while Aristotle held that life could spontaneously generate from inanimate matter. But Plato's geometrical vision of the structure of matter has definite echoes in the concepts of modern theoretical physics. In his intuition we can now see a deep truth about the fundamental make-up of the universe.

Plato recorded his philosophy in the form of dialogues in which his characters present their philosophical ideas and argue their case with one another. These fictionalised dialogues present education as a form of theatre, in which the students arrive at the truth through conversation and debate, guided by the hand of their teacher, in this case Socrates. One such dialogue, the *Timaeus*, opens with Socrates gathering the participants together before delving into a discussion of the structure of matter:

> SOCRATES: One, two, three – but where, my dear Timaeus, is the fourth of my guests of yesterday who were to entertain me today?

TIMAEUS: He's fallen sick, Socrates; otherwise he would never willingly have missed today's discussion.

SOCRATES: Then if he's away it is up to you and the others to play your part as well as his own.[6]

As the dialogue develops, Plato offers, through the mouth of Timaeus, a description of how geometry might manifest itself in the structure of matter. It is the oldest geometrical theory of matter we know of. This model has no basis in reality, and Plato may not have believed that it was literally true, as he offered no physical evidence to support it. He presented many of his ideas as allegories, so it could be that the model in the *Timaeus* was intended to be understood in this way. Even so, it was an illustration of how elegant geometrical facts might be used to explain the structure of matter, and as such it was a stimulus to further investigation.[7] It would certainly prove to be an influential vision of how geometry could be used to model the laws of nature.

Puzzle 1

Take six matches and arrange them in such a way that they form four triangles, without any of the matches crossing over one another.

The theory of matter described in the *Timaeus* builds on the theory originally proposed by an earlier philosopher, Empedocles, that all matter is composed of combinations of four elements[8]: earth, air, fire and water. This idea probably arose to account for the transformations of matter in familiar everyday physical processes. For instance, a block of ice is solid, but if heat is applied to it then the ice will melt. This might be interpreted as taking a solid (earth) and adding heat (fire) to produce a liquid (water): earth plus fire produces water. Similarly, if the water is heated it will evaporate, so water plus fire produces air.

But Plato, through Timaeus, goes further. He proposes what could be considered as a subatomic theory of the four elements. He suggests that, at the most fundamental level, each element is made up of tiny components, each with the shape of a regular solid. For this reason, the regular solids are often known as the Platonic solids. (A regular solid, or regular polyhedron, is a three-dimensional shape whose faces are all regular polygons of the same type, with the same number of polygons meeting at each of the corners, or vertices, of the solid.[9] For instance, a cube has six square faces, with three squares meeting at each corner. The cube is therefore a regular solid.)

There are just five different regular solids, as shown in Figure 1, and this is the basis of Plato's scheme. According to him, earth is formed of cubes, fire is formed of tetrahedra, air is formed of octahedra and water is formed of icosahedra. Part of the motivation for this model seems to be the idea that the triangular faces of the tetrahedra, octahedra and icosahedra could be rearranged, like subatomic units.

Answer to Puzzle 1

It isn't possible to connect the six matches to form four triangles without any of them crossing over another one, if the matches are restricted to lying in a plane. However, if the third dimension is used, the matches can be positioned to form the edges of a tetrahedron. The tetrahedron has six edges, which enclose four triangular faces, thus solving the puzzle.

Figure 1 Skeletons of the regular solids in which only their edges are shown. From left to right: tetrahedron, octahedron, icosahedron, cube, dodecahedron.

This would enable the fundamental tetrahedra, octahedra and icosahedra to transmute into one another, and thereby explain the transformations of matter. For example, a tetrahedron has four triangular faces and an octahedron has eight triangular faces. So in Plato's model two tetrahedra could be converted into a single octahedron,[10] and this would represent an alchemical transformation of fire into air (see Figure 2). Unlike terrestrial matter, which was subject to endless transformation and decay, Plato considered the heavens to be eternal and unchanging. Celestial matter, in Plato's view, was therefore composed of a single, perfect fifth

Figure 2 The five Platonic solids represented as the elements by the German physicist Johannes Kepler. From left to right, the top row shows: the octahedron, the tetrahedron and the dodecahedron. On the bottom row are the cube and the icosahedron.

element. This celestial element later became known as quintessence. In Plato's scheme in the *Timaeus*, the heavens were associated with the fifth regular solid, the dodecahedron.

Kepler and Tycho

Plato's model certainly does not correspond to the nature of reality, but it does demonstrate a curious intuition about the structure of matter. Over the course of the twentieth century, physicists increasingly built their theories on geometry and symmetry, and with great success, so Plato's surprising idea is one that resonates with modern physicists. Furthermore, his geometrical alchemy had a great influence on Renaissance thinkers, so in a sense the geometrical theories of today are the great-grandchildren of this model. In particular, it provided lifelong inspiration for the German mathematician and astronomer Johannes Kepler (1571–1630), who straddled the divide between the medieval age of superstition and the start of the modern scientific era. Kepler was captivated by symmetry and patterns, and, like Plato, he believed that God had constructed the universe in accordance with elegant geometrical principles.[11]

Kepler had an unruly upbringing in a tavern in the town of Weil-der-Stadt in western Germany. When he was still a young child, his dissolute father abandoned his family to lead the life of a mercenary soldier. Much later in his life, Kepler's mother was imprisoned and tried for witchcraft, and was released only after her health had been ruined. Throughout his eventful career Kepler was caught up in the religious turmoil of the Thirty Years War that raged across Central Europe. Amid this chaos, he continued to search for the harmony in the heavens. Kepler was one of the very first modern scientists. He made great contributions to mathematics and the science of optics, but he is most famous for his astronomical research, not as an observer – he suffered from poor eyesight – but as a theorist.

One of the most decisive partnerships in the history of science was that between Kepler and the irascible Danish nobleman Tycho Brahe (1546–1601), who had dedicated his life to charting the night sky and setting astronomy on a secure footing. As a student, Tycho lost the bridge of his nose in a duel and thereafter hid his wound with a prosthetic metal nose held in place with a sticky paste. During his student days, Tycho realised that all the existing astronomical almanacs predicting the positions of the planets in the night sky were wildly inaccurate, and he drew up a long-term plan to amend them. The Danish king approved of the task and granted Tycho the island of Hven, where he could build an observatory and start his astronomical research. The haughty astronomer ruled the island as his own personal fiefdom and maintained a court there that included his jester dwarf and his pet giant elk. Tycho spent over two decades designing and constructing enormous astronomical instruments which he used to make the most accurate set of observations of the stars and planets before the invention of the telescope. But the king died in 1596, and his successor did not take the same interest in his science. He moderated Tycho's lavish expenditure, and it was not long before they fell out irreconcilably. In high dudgeon, Tycho left Hven along with his entire household and entourage, many of his gargantuan instruments and his trunk full of meticulous observations, to offer his services where they might be better appreciated – to the Holy Roman Emperor Rudolf II in Prague.

Tycho had chosen his new patron well, and was duly appointed Imperial Mathematician to the Holy Roman Empire. On the first day of January in the year 1600, Johannes Kepler set off to meet Tycho and take up the role of his assistant. This proved to be the greatest conjunction in the history of astronomy. It lasted for just over a year. In 1601 Tycho died suddenly as the result of a drinking binge, and Kepler soon became the new Imperial Mathematician. More importantly, he also took possession of Tycho's records, tracking the exact positions of the planets across the night sky over the course of many years. Kepler's painstaking analysis of these observations would instigate a scientific revolution.

True to their devotion to regular geometrical figures, the ancient Greek mathematicians had insisted that the planets could move only in orbits that were perfect circles. For over two thousand years, astronomers had persisted with the notion that the planets moved around circles on circles called epicycles. This system, which had been passed down to Western astronomers in an Arabic translation of the work of the Greek scholar Ptolemy, had become ever more elaborate with age as additional circles were incorporated to keep the theory in line with observations.[12] Throughout the centuries, no one had been brave enough to challenge the authority of the ancients, but Kepler knew that he had the best collection of observations that had ever been amassed, and he was determined to find the correct model of the heavens. Like the Greek philosophers, Kepler was a great believer in beautiful geometry, but unlike his predecessors he would not accept any theory that did not tally with accurate observations.

After many years of complicated and tedious calculations, Kepler ruled out epicycles and various other figures as the correct shape of the planetary orbits and finally showed that Tycho's observations could be explained only if the planets followed elliptical paths around the Sun (see Figure 3). (An ellipse is a squashed circle, or equivalently the shape we see if we view a circle at an oblique angle.) This was an enormous conceptual revolution, and it marks the beginning of modern astronomy. Along with the other laws of planetary motion that Kepler deduced from Tycho's observations, this discovery played a major role in enabling Newton to formulate his laws of motion and his universal law of gravitation.

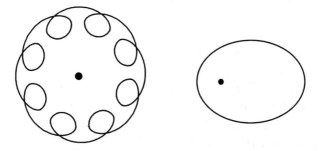

Figure 3 Left: The epicyclic path of a planet around the Earth in the classical Greek model. Right: The elliptical path of a planet around the Sun in Kepler's model. The eccentricity of the ellipse is exaggerated to illustrate how it differs from a circle; the orbits of the planets in the solar system are much closer to circles than this.

Snowflakes Are Dancing

In 1611, Kepler received an invitation to join the New Year's celebrations of his friend Johannes Matthäus Wackher von Wackenfels, privy counsellor to the Emperor. Kepler later recalled how, one winter's day while he was musing over an appropriate New Year's gift for his friend, by a happy chance water-vapour was condensed by the cold into snow, and specks of down fell here and there on my coat, all with six corners and feathered radii. Upon my word, here was something smaller than any drop, yet with a pattern; here was the ideal New Year's gift ... the very thing for a mathematician to give ... since it comes down from heaven and looks like a star.[13]

Unable to preserve a beautiful and intricate snowflake as his present, Kepler decided to write a little booklet about the snowflake and its symmetry in honour of his friend. This booklet is called *De Niva Sexangular* ('The Six-Cornered Snowflake'). Kepler realised that although the exact shape of the icy filigree of each snowflake is different, they all display the same hexagonal symmetry. It was the origin of this symmetry that intrigued him. In the booklet, he looked for clues in the geometrical structure of other familiar symmetrical objects such as crystals, bees' honeycombs and the seed cases of pomegranates. Kepler was aware of the speculations of the ancient philosophers on the fundamental atomic structure of matter, and wondered whether it was the geometrical arrangement of the atoms in the ice that is ultimately responsible for the symmetry of the snowflakes.

Atoms

If, in some cataclysm, all of scientific knowledge were to be destroyed, and only one sentence passed on to the next generations of creatures, what statement would contain the most information in the fewest words? I believe it is the *atomic hypothesis (or the atomic fact, or whatever you wish to call it), that all things are made of atoms – little particles that*

move around in perpetual motion, attracting each other when
they are a little distance apart, but repelling upon being squeezed
into one another. In that one sentence, you will see, there is
an enormous amount of information about the world, if just
a little imagination and thinking are applied.

Richard Feynman, *The Feynman Lectures on Physics* (1963)[14]

The idea that matter is constructed from minuscule
entities called atoms dates back to the Greek philosophers of
the fifth century BCE. This hypothesis was the outcome of a
purely philosophical debate about whether it was possible to
continue slicing matter indefinitely into ever smaller pieces.
Democritus believed that this process would have to come to an

---

Puzzle 2

Given a collection of equal-sized circles, what is the largest
number of circles that each circle can touch without any of
the circles overlapping? Mathematicians call this number the
kissing number of the circle. Try it with equal-sized coins.

---

end when the fundamental components of matter were reached.
He called these ultimate constituents of matter 'atoms', a word
that literally means 'indivisible'. Atomism went in and out of
fashion over the centuries, but as long as the debate about atoms
was based solely on philosophical arguments, it could never be
conclusively decided one way or the other.

Kepler took the idea of atoms and decided to represent
them in the simplest way mathematically as tiny, equal-sized
spheres. He assumed that in solids the atoms are packed
together as closely as possible, and set himself the task of
exploring the different ways in which they could be arranged.

Answer to Puzzle 2

The maximum kissing number of a circle is 6, as achieved in the packing shown in Figure 4.

If he could find the densest packing of spheres – the arrangement that packed the maximum number of spheres into a given volume – he thought it might shed some light on the origin of the structure of crystals.

Kepler began by considering the equivalent problem in two dimensions, which is much simpler. How do you get the maximum number of equal-sized circles in a given area? His solution to this problem is shown in Figure 4. This arrangement is rigid – it does not allow any wriggle-room. If one circle is going to move, then they will all have to move. It is clear that this must be the densest packing for circles, and this produces the solution to Puzzle 2: each circle is touched by six neighbouring circles. So when the circles are packed this way, the kissing number of each circle is 6. That is the highest kissing number that a circle can have. No more circles can be packed around the central circle, which is easy to demonstrate with seven coins of the same denomination. Six of the coins will fit exactly around the seventh.[15]

Having cracked the simpler two-dimensional packing problem, Kepler returned to his original three-dimensional problem. The packing he thought must be the densest was the one you can see in shops today in which oranges or other similarly shaped fruit are stacked in pyramids. Each layer is arranged like the two-dimensional packing shown in Figure 4. In Kepler's packing, each layer is stacked as depicted in Figure 5, which shows the positions of three successive layers. Each layer of circles represents the 'equators' of a layer

Figure 4 The densest packing of circles in a plane.

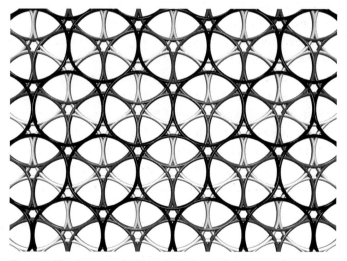

Figure 5 The three sets of differently shaded circles represent the equators of three successive layers of spheres in the densest possible packing of spheres in three-dimensional space.

of spheres. We can label these three layers A, B and C, from bottom to top. In Figure 5, the next layer up would be positioned so that its spheres are directly above those in layer A, the following one up with its spheres directly above those in layer B, the one above that with its spheres directly above those in layer C, and so on. Kepler thought that this packing must be the densest possible, but he was unable to find a conclusive proof. The kissing number of the spheres in Kepler's packing is 12. The left side of Figure 6 is an exploded view that shows that each sphere in this packing arrangement is surrounded by a hexagon of six spheres in a plane; a triangle of three other spheres rests on top of the central sphere, and another triangle of three spheres sits below it. The centres of the twelve spheres that touch the central sphere are situated at the vertices of a polyhedron known as a cuboctahedron.[16]

Figure 6 Left: This exploded view of the twelve spheres surrounding a central sphere shows that the kissing number of a sphere in Kepler's packing is 12. Right: The centres of the twelve spheres surrounding a central sphere in Kepler's packing are situated at the vertices of a cuboctahedron.

Kepler's idea that the atoms in solids are arranged like closely packed spheres has proved to be remarkably astute. It turns out that when equal-sized atoms form a crystal, they will often arrange themselves in Kepler's packing.[17] The structure of atoms is now well understood. They are not really perfect spheres, like tiny ball bearings, so there are other factors that influence the way they will stack together to form crystals. However, although various other arrangements are possible, the atoms in many metals, such as copper, silver and gold, do indeed organise themselves in just the way that Kepler envisaged. It is quite incredible that with no more inspiration than a falling snowflake, Johannes Kepler was able to muse on the fundamental structure of matter and draw such a deep and basically correct conclusion.

As no one since Kepler has found a denser packing than his, most chemists and physicists are probably happy to accept that this packing must be the densest possible. But for mathematicians there is always room for doubt. They demand a cast-iron, irrefutable mathematical proof. Further storeys can be added to the edifice that is mathematics only after every single brick has been shown to be flawless and unbreakable. But Kepler was unable to find a rigorous mathematical proof, and neither could the mathematicians who followed him. His problem was to become one of the oldest outstanding problems in mathematics. It became known as the sphere-packing problem, or Kepler's conjecture, and it resisted the ingenuity of mathematicians for almost four hundred years.[18] A rigorous proof eluded everyone until 1998, when Thomas Hales of the University of Pittsburgh announced that he had finally succeeded.[19] Unfortunately, Hales's proof does not fall into the category of ideal mathematical proofs sought by Euclid and his colleagues. An elegant proof in the style of the classical Greek geometers would fit on the back of an envelope and provide stimulating new insights into geometry. Hales's proof consists of around 250 pages of logic and culminates in an extremely complicated equation whose analysis requires the use of a computer, so the proof also includes 3 GB of data generated by his computer analysis – the equivalent of about four CD-ROMs' worth of data. This makes the proof very difficult for other mathematicians to evaluate, but the consensus seems to be that the proof is correct.[20]

It was many years after Kepler's time that the existence of atoms became generally accepted. The case for atoms was greatly strengthened when scientists realised that substances might be composed of a variety of atoms, and that the differences between the atoms could explain their differing chemical properties. Chemistry as a subject was, however, still in its infancy in Kepler's day. There were plenty of alchemists in Europe, and some, such as Tycho, were serious scientific researchers. But many others were tricksters who defrauded greedy kings and emperors out of large sums of money in exchange for the promise of unlimited wealth when they had perfected their technique for transforming base metal into gold. No doubt these alchemists were adept at performing strange chemical transformations. The patrons of the alchemists appear incredibly gullible to us today, but the idea of transmutation of the elements is not quite as crazy as it might at first seem.

Chemical transformations can be very surprising indeed. For example, sodium is a soft, silvery-grey metal that reacts with great intensity when placed in water. It will fizz around violently and become very hot. Chlorine is a poisonous, greenish gas. Yet, if sodium and chlorine react together, they will combine to form a shiny white crystalline solid that we know very well and eat regularly as part of a normal diet. This compound, sodium chloride, is table salt. Unlike sodium, it dissolves in water without any fuss or histrionics. Its properties are completely different from the properties of its two components. If a soft grey metal and a toxic green gas will react to form table salt, then maybe it was not so ridiculous for the alchemists to imagine that they could make gold or any other material if they went about it in the right way.

The Chemical Elements

> I now mean by elements, as those chymists that speak plainest do by their principles, certain primitive and simple, or perfectly unmingled bodies; which not being made of any other bodies, or of one another, are the ingredients of which all those called perfectly mixt bodies are immediately compounded, and into which they are ultimately resolved.
>
> Robert Boyle, *The Sceptical Chymist* (1680)[21]

The evolution of alchemy into a modern science was a long and gradual process. It was not until the Age of Enlightenment in the eighteenth century that the arcane art metamorphosed into a precise quantitative science. The person who was most important in this modernisation process was Antoine-Laurent Lavoisier (1743–94), who is often regarded as the father of chemistry. Lavoisier, born into a wealthy bourgeois French family, greatly increased his personal fortune by accepting a position as a tax collector, but his passion was chemistry. He used his wealth to build a well-equipped laboratory where he could advance his chemical investigations.

Most of Lavoisier's research was conducted in collaboration with his wife, Marie-Anne, who acted as his assistant, writing up the notes of the experiments and making detailed and accurate drawings of the apparatus. The couple were close friends of the artist Jacques-Louis David, who painted their portrait in his grand neoclassical style (see Figure 7). In the painting, a devoted Marie-Anne leans against her seated husband, who looks up at her, quill in hand, no doubt recording some important observation. On the table before them, and on the floor in the foreground, are some of the glass

Figure 7 Portrait of Antoine-Laurent and Marie-Anne Lavoisier by Jacques-Louis David (1788).

instruments used in their scientific research. Lavoisier purchased the finest balances that were available in his day. He could measure mass to an accuracy of 1 part in 600,000.[22] The great precision he achieved with his measurements enabled him to take great strides in the understanding of chemical reactions and turn his subject into an accurate quantitative science. His aim was to transform chemistry into a modern subject and put it on a firm theoretical footing.

Lavoisier found that while most materials could be chemically decomposed into simpler substances, some chemicals appeared to be fundamental in that they could not be broken down any further. These fundamental substances are the chemical elements. Lavoisier drew up a list of thirty-seven substances he considered to be elementary. In his *Elementary Treatise on Chemistry*, published in 1789, Lavoisier demolished the ancient theory of the four elements (earth, air, fire and water) and replaced it with a theory of the chemical elements that is still recognisable today. He demonstrated that water could be decomposed into hydrogen and oxygen; that air was composed of a number of different gases; and, as for the ancient earth element, he succeeded in isolating several 'earths' that he was unable to decompose any further. It is now clear that a couple of the 'elements' in Lavoisier's list, such as light and heat, are not chemical substances at all. There are also a few that further analysis has shown to be compounds, but chemists today still recognise most of his entries as chemical elements. These include the metals copper, gold, lead, iron, mercury, silver, tin and zinc, and the non-metals sulphur, phosphorus, oxygen, hydrogen, nitrogen and carbon.

Lavoisier's remarkable scientific research was cut tragically short. During the Reign of Terror that gripped Paris after the French Revolution, Lavoisier was arrested along with a number of other tax collectors and, after a hasty trial, found guilty of treason. On 9 May 1794 he was taken to the Place de la Revolution and guillotined. The mathematician Joseph-Louis Lagrange exclaimed sorrowfully that 'It took them only an instant to cut off that head and a hundred years may not produce another like it'.[23]

John Dalton

'Did you never study atomics when you were a lad?' asked the Sergeant, giving me a look of great inquiry and surprise.

'No,' I answered.

'That is a very serious defalcation,' he said, 'but all the same I will tell you the size of it. Everything is composed of small particles of itself and they are flying around in concentric circles and arcs and segments and innumerable other geometrical figures too numerous to mention collectively, never standing still or resting but spinning away and darting hither and thither and back again, all the time on the go. These diminutive gentlemen are called atoms. Do you follow me intelligently?'

'Yes'.

'They are lively as twenty leprechauns doing a jig on top of a tombstone.'

Flann O'Brien, *The Third Policeman* (1940)[24]

Chemists remember John Dalton as the originator of the theory that chemistry can be explained in terms of the atomic constituents of matter. Dalton was born in the small Cumbrian village of Eaglesfield in 1766.[25] His family were Quakers, and his father earned a meagre income by working as a weaver. Education was highly valued in the Quaker community, and

John was provided with as much instruction as was available in their rural location. By the age of fifteen he was teaching at a boarding school in Kendal. In this small Lakeland town he met a local intellectual, John Gough, who despite being blinded by smallpox at the age of two, had learnt Latin, Greek and French and had a good knowledge of natural philosophy, as science was termed in those days.

Gough took the young Dalton under his wing and tutored him in classical languages and natural philosophy. He also gave Dalton access to his library and his scientific instruments, and encouraged his interest in meteorology. Dalton would keep a meteorological diary throughout his life. In 1792, Gough recommended Dalton for a teaching position at New College in Manchester. The college duly offered the post to Dalton, who accepted it without hesitation. The following year he was elected to the Manchester Literary and Philosophical Society, an organisation that would play an important role in his future life. During the next fifty years he presented over a hundred papers to the society.

Dalton's greatest contribution to science was to show how chemical processes could be better understood by accepting the idea that matter is composed of atoms. His work built on that of Lavoisier, who had argued that the foundations chemistry should be the properties of a number of simple chemical substances that could not be further broken down – the chemical elements. Dalton proposed that the differing chemical properties of the elements could be explained if each element is assumed to be composed of its own unique type of atom (see Figure 8). With a different atom associated with each element, the vast array of known chemical compounds would then be formed by combining atoms together in various ways to form compound atoms – or in modern terminology, molecules.

Dalton's investigations convinced him that when chemical elements combine, they always do so in the same proportions. For instance, 1 gram of hydrogen would combine with 8 grams of oxygen to form 9 grams of water; 2 grams of hydrogen would combine with 16 grams of oxygen to form 18 grams of water; and so on. Dalton believed that this could be explained if hydrogen gas were formed of hydrogen atoms and oxygen gas were formed of oxygen atoms, and that these two types of atoms always combined in the same way to form a compound atom of water. He assumed that the compound atom would be formed from one hydrogen atom and one oxygen atom. Today we would call the compound atom a molecule of water. He deduced from his experiments that the oxygen atom must have seven times the mass of a hydrogen atom.

Dalton's idea was essentially correct, but for a couple of reasons the values he derived for the relative masses were inaccurate. First, Dalton's measurements were not very precise. His table of atomic masses, shown in Figure 8, gives the atomic mass of hydrogen as 1, but the atomic mass of oxygen as 7, not 8 – the value he would have obtained if his measurements had been accurate. The second and more fundamental reason that Dalton's values were incorrect is that he was unable to deduce the number of atoms in a molecule. We now know that a water molecule is formed from two hydrogen atoms and one oxygen atom, so a water molecule is represented by the chemical formula H_2O. With the correct formula, it can be deduced that an oxygen atom actually has sixteen times the mass of a hydrogen atom and not eight.

Establishing the correct formulae for the atomic constituents of molecules requires another piece of information. The vital clue came when it was discovered that gases always combine in regular proportions of their volumes. For example, two

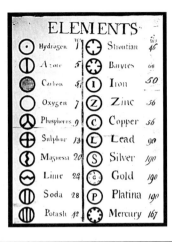

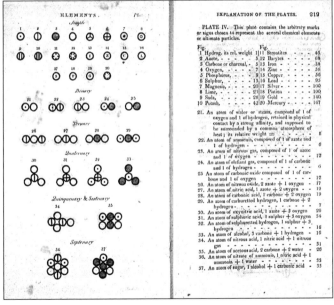

Figure 8 Above: John Dalton's table of atomic masses. Below: His diagrams of 'compound atoms' (molecules).

litres of hydrogen will combine with one litre of oxygen to form two litres of water vapour. The Italian chemist Amedeo Avogadro hypothesised that equal volumes of gas contain equal numbers of molecules, and this enabled him to deduce the correct form of chemical equations such as

$$2H_2 + O_2 \rightarrow 2H_2O$$

where H_2 represents a hydrogen molecule, which is formed of two hydrogen atoms; O_2 represents an oxygen molecule, which is formed of two oxygen atoms; and H_2O represents a molecule of water.

By the middle of the nineteenth century, all the pieces of the chemical jigsaw were available, but scientists still needed to identify the main features of the picture and assemble all the pieces. For this purpose, an international conference was convened at the University of Karlsruhe in Germany to thrash out the differences and establish the correct atomic masses and chemical formulae. At this conference, which took place in 1860 and was the first of its kind, the chemical and physical properties of all the known elements were placed on a secure foundation. All the information was now available for a grand synthesis of the subject of chemistry.

Chemical Patience

Portraits of Dmitri Mendeleyev (Figure 9) give us the impression of a long-haired and bearded Russian orthodox anchorite, bringing to mind a character from a Dostoyevsky novel or even the 'mad monk' Rasputin. But Mendeleyev is the world's most famous chemist and the author of the periodic table of the elements, the key to the whole subject of chemistry.

Mendeleyev was born in 1834 in Tobolsk, a small town in western Siberia, the youngest in a family of fourteen children.[26] His father was a local schoolteacher who became blind in the year of Dmitri's birth. Providing for the huge family then became the responsibility of Dmitri's remarkable mother, Maria. Maria's father had established a paper factory, a printing press and a glass factory in the area, and she

Figure 9 Dmitri Mendeleyev.

was forced to reopen the glassworks in order to generate an income to support the family. Dmitri's father died, and the ill-fortune of the family was compounded the following year when the glassworks burnt to the ground.

Despite their hardships and misfortunes, Maria was determined to obtain the best possible education for her brilliant son Dmitri. In 1849 she set out with the fifteen-year-old boy and his sister Liza on an epic journey of over two thousand kilometres, from village to village on horse-drawn wagons, across the bleak plains of Russia to the great city of Moscow. The qualifications Dmitri had obtained at his school in their provincial Siberian town did not impress the officials in Moscow, and Maria was unable to persuade them to allow him to continue his education there. Her spirit undaunted, Maria set off with Dmitri and Liza to travel a further 650 kilometres north to the capital of the Russian Empire, St Petersburg. Within weeks of obtaining a government scholarship for Dmitri to study mathematics and natural sciences at the Main Pedagogical Institute in St Petersburg, the exhausted Maria died.

Maria's fatal determination to obtain the best possible university education for her son would pay great dividends. Mendeleyev completed his studies in 1855 and earned a gold medal as the best student in his year. Four years later he was awarded a grant for advanced study abroad, and on the advice of his friend, the chemist and composer Alexander Borodin, he left Russia to work for his doctorate in Paris. Mendeleyev continued his studies in Heidelberg under the German chemist Robert Bunsen before returning to St Petersburg in 1861. On his return, Mendeleyev was appointed professor of chemistry at the University of St Petersburg and began writing a textbook, *The Principles of Chemistry*, that would become a classic, with translations into English, French and German. While planning the outline of the book, Mendeleyev realised that he needed to find a way to organise the chemical elements that would illustrate the connections between their chemical properties and provide a natural framework for the material he was compiling. He brought his encyclopaedic knowledge of chemistry to bear on this problem. His solution would change chemistry for ever.

Early one morning in 1865, while he was waiting for a sleigh to take him across the snow to the railway station, Mendeleyev picked up a pack of playing cards to pass a few quiet moments playing patience.[27] He shuffled the pack, dealt the cards and began to play. At the end of his game the cards were arrayed before him with each suit – clubs, diamonds, hearts and spades – in its own row, and the card values – king, queen, jack, ten, and so on – ordered in columns. Looking at this pattern, Mendeleyev had a brilliant idea: perhaps the chemical elements could be organised in a similar way to the playing cards. Maybe there was a similar pattern to the elements. He made a set of blank cards and on each one he wrote the symbol for one of the chemical elements. He then began moving the cards around to see what patterns he could make based on the atomic masses of the elements and their chemical properties. The sleigh arrived to take him to the station, but he ignored it and continued to search for a pattern. He struggled through the morning, playing what he would later call his 'chemical patience'. He couldn't quite make sense of the patterns, but he knew he was very close to an important breakthrough. Eventually he asked his housekeeper to dismiss the sleigh-driver with a request to return in time for Mendeleyev to catch the afternoon train. Finally, frustrated and drowsy, he fell asleep in his chair. It was during this brief nap that the structure of the periodic table came to him: 'I saw in a dream a table where all the elements fell into place as required,' he later recalled. 'Awakening, I immediately wrote it down on a piece of paper.'[28]

When placed in order of their atomic mass, the elements show a regular pattern – a periodicity – in their chemical properties. The modern periodic table is shown in Figure 10. Comparing elements 9, 10 and 11 with the elements eight places further on in the table, namely elements 17, 18 and 19, will give a quick illustration of the periodicity upon which the table is built. Element number 9, fluorine (F), is a poisonous and extremely reactive gas. Moving on eight places to element number 17, which is positioned below fluorine in the table, we come to chlorine (Cl) – which is also a poisonous and very reactive gas. Element number 10

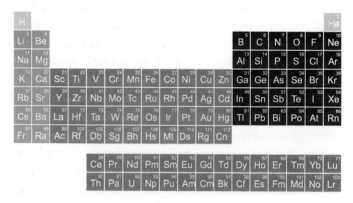

Figure 10 The modern periodic table of the elements.

in the final column of the table is neon (Ne), which is a completely unreactive or inert gas. Eight places further on and beneath neon lies element number 18, argon (Ar) – an inert gas. Element number 11, sodium (Na), is a very reactive soft grey metal. Moving on eight places brings us to element number 19, potassium (K), just beneath sodium in the first column of the table. Potassium is also a very reactive soft grey metal.

The periodic table is constructed so that the columns are formed by lining up elements with similar properties. Typically, the chemical properties of the elements in a column will show a trend as we move down the column. For instance, the elements in the first column react vigorously with water, but as we move down the column – from lithium (Li) to sodium to potassium, and so on – this reactivity dramatically increases. If the elements are the iridescent silken threads from which the rich tapestry of chemistry is woven, then the periodic table is the universal pattern book containing all the subtly coloured threads arrayed in perfect order.

Other nineteenth-century chemists had recognised that there was some regularity in the properties of the elements, but none found a systematic and useful way to organise them. Mendeleyev was convinced that his periodic table reflected the true nature of the chemical elements and was not just some arbitrary construction that his imagination had imposed on them. The strength of his conviction was such that he explained away apparent discrepancies in the table by arguing that some of the published measurements of atomic masses might be incorrect. He also suggested how the gaps in his table could be plugged, by predicting the existence of previously unknown elements.[29] Below aluminium (Al), for instance, there was a gap in his table. Mendeleyev was convinced that there must be a missing element that would fill the gap. He deduced that the atomic mass of this element should be about 68 and calculated its density, boiling point and the properties of various chemical compounds that the element would form. He named the element 'eka-aluminium', meaning 'the one after aluminium' (*eka* means 'one' in Sanskrit).

A new element closely matching Mendeleyev's predictions for eka-aluminium was isolated by the Frenchman Paul

Émile Lecoq de Boisbaudran in 1875, and in honour of his native country he named it gallium (Ga), from the Latin name for France. Mendeleyev was delighted when he heard the news. However, when de Boisbaudran measured the density of gallium it appeared to be significantly lower than Mendeleyev had predicted. Unperturbed, Mendeleyev suggested that the Frenchman should check his measurement. The Russian was vindicated: the new measurement was very close to his prediction. Four years later he had another success. The Swedish chemist Lars Nilson identified an element that corresponded to Mendeleyev's eka-boron. Nilson named the element scandium (Sc), after Scandinavia. Then in 1886 the German Clemens Winkler discovered a new element corresponding to the element eka-silicon, predicted by Mendeleyev to fill a gap below silicon. It probably didn't take a chemist with the insight of Mendeleyev to predict that Winkler would follow the trend of recent discoveries and name the new element germanium (Ge) after his homeland.[30]

The version of Mendeleyev's table that he published in 1871 is essentially the periodic table that chemists still use today, although a couple of significant amendments were made later. The most important of these was the incorporation of the inert gases: helium (He), neon (Ne), argon (Ar), krypton (Kr), xenon (Xe) and radon (Rn). They are all chemically unreactive gases (though in the past fifty years compounds of some of them have been produced). These colourless and odourless gases were unknown when Mendeleyev originally devised his table, and were discovered in the last years of the nineteenth century. The invention of the periodic table brought a coherent order to the science of chemistry, and it remains the cornerstone of the subject. It allows chemical reactions and the properties of the elements to be explained in terms of their constituent atoms.

Ultimately, these properties follow directly from the physical composition of the atoms. The exact nature of this relationship would not be established until the structure of atoms was understood, in the 1920s.

Plato's solids, Kepler's snowflakes and Mendeleyev's chemical periodicities all offer tantalising glimpses of the patterns at the heart of matter – and, as will be revealed, symmetry is the key to understanding the structure of matter at every level. In the 1960s, the protons and neutrons from which atoms are composed turned out to be formed from even more fundamental entities, known as quarks. The existence of quarks was originally proposed when it became clear that the new particles produced in accelerator experiments fall naturally into patterns that hint at a deeper nuclear substructure. Today, all the known fundamental particles, such as the electron and the various quarks, can be organised into a table of the elementary particles, which has echoes of the periodic table and represents order at a deeper level. As with the periodic table, there were empty slots in this new table, but in recent years they have all gradually been filled as new fundamental particles were discovered. The final gap in the table has now been plugged by the discovery of the last elusive particle – the Higgs.

Symmetry also plays a vital role in modern theories of the forces that hold matter together, but to unlock the secrets of this enchanted realm of modern physics requires a more sophisticated key than any available to those earlier researchers. Fortunately, mathematicians have developed the perfect precision instrument for the task which gives them a systematic understanding of symmetry and opens the door to particle physics. This is the subject of the following sections.

It's All Done with Mirrors

> Fate is partial to repetitions, variations, symmetries.
>
> Jorge Luis Borges, 'The Plot' (1957)[31]

David Brewster was born on 11 December 1781 in the town of Jedburgh in Scotland. He entered the University of Edinburgh as a twelve-year-old child prodigy and would become one of the great British scientists of the nineteenth century. Brewster constructed his first telescope at the age of ten and retained a lifelong interest in optics and scientific instrument-making. Throughout his long career he invented many ingenious optical devices, but the instrument he is most fondly remembered for is a toy. In 1816, Brewster, by now a renowned scientist, constructed the world's first kaleidoscope, a device that is just as popular today as it was two hundred years ago. Its only function is to dazzle the eye with an endlessly changing sequence of symmetrical images. Brewster named his invention the 'kaleidoscope', combining three Greek words: *kalos*, meaning 'beautiful', *eidos*, meaning 'shape', and *scopos*, meaning 'viewer' – the viewer of beautiful forms.

The kaleidoscope is an enclosed tube containing a number of precisely aligned mirrors. At one end of the tube is a compartment containing coloured glass beads. This end, the objective end, is translucent to allow light into the tube to illuminate the beads. At the other end of the tube is an aperture through which you can peer into the tube. What you see is a beautiful symmetrical pattern produced by the multiple reflections of the beads in the mirrors. Twisting the tube shuffles the beads around so that the symmetrical patterns morph into a never-ending sequence of new patterns. It is a simple idea, but one with a timeless charm.

Brewster applied for a patent for his invention, but unfortunately the application was rejected on the grounds that he had already demonstrated in a public lecture how the kaleidoscope worked. The kaleidoscope was an immediate commercial success and sold in great quantities: 200,000 were reported to have been sold in London and Paris in just three months.[32] However, Brewster's simple design was easily copied, and nearly all the kaleidoscopes that were purchased were cheap imitations that did not work as well as Brewster's original. While others made a fortune marketing his invention, Brewster received little financial reward for it.

It is the arrangement of the mirrors inside the kaleidoscope that is the source of its magic. Brewster's kaleidoscope contains two rectangular mirrors running lengthways down the tube. The mirrors are joined along one of their long edges and fixed at an angle to each other to give a V-shaped cross section, with their mirrored sides facing inwards. The angle between the faces of the two mirrors is 30 degrees, or exactly $\frac{1}{12}$ of a full rotation. When an assortment of coloured beads

---

Puzzle 3

What is the next number in this sequence?

1, 8, 11, 69, 96, 101, 111, 181 ...

---

is placed between the mirrors, the entire collection undergoes multiple reflections and we see twelve copies of the beads, six of which are mirror-reversed. The image produced by the kaleidoscope is a hexagonal, psychedelic snowflake-like pattern. As the tube is rotated, the beads rearrange themselves and the pattern changes, but always retains the same hexagonal symmetry.

The upper image in Plate 1 was generated by constructing a virtual computer model of Brewster's kaleidoscope. At the objective end of the kaleidoscope an intricate fractal design was projected. The pretty snowflake-like pattern with its hexagonal symmetry is the result of multiple reflections of this design in the two mirrors of the virtual kaleidoscope. Most kaleidoscopes sold in toyshops today contain three mirrors that are joined together so that they have a triangular cross section. This gives the kaleidoscopic patterns a higher degree of symmetry, like the image shown in the lower half of Plate 1, which was produced with a virtual kaleidoscope of this type. Here the multiple reflections of the pattern have produced an image that will repeat endlessly in all directions to cover a flat two-dimensional surface. The symmetry of this pattern is the same as the symmetry of the densest packing of circles described by Kepler in his little booklet about the snowflake.

Answer to Puzzle 3

The numerical representations of these numbers are all symmetrical under 180-degree rotations around their centre. The next number in the sequence is 609.

To Infinity and Beyond

We all have an intuitive feeling for what symmetry means. We recognise symmetrical patterns when we see them, such as the kaleidoscopic images shown here. But mathematicians are not content with simply recognising symmetry when it shows up. They demand a clear definition of what symmetry means in its most fundamental sense, as it is only by building on precise, unambiguous definitions that mathematical certainty is possible. The symmetry of simple figures such as a snowflake or Platonic solid may be easy to visualise and understand, but a more formal approach to symmetry is required if mathematicians are to make sense of objects with an even higher degree of symmetry.

The mathematical analysis of symmetry dates back to investigations made nearly two centuries ago by a Norwegian, Niels Abel, and a Frenchman, Évariste Galois.[33] These two mathematicians both died pitifully young. In 1829, at the age of just twenty-six, Abel succumbed to tuberculosis, contracted while living in poverty as a student in Paris. Tragically, he died just as his mathematical talent was about to be recognised with the offer of a professorship. Galois famously died in 1832 at the age of only twenty after being shot in a duel, having spent the whole of the previous night frantically scribbling down the results of his mathematical researches. Galois's work remained unknown until his manuscript was edited and published fourteen years later by another French mathematician. But the first thorough exposition of the mathematics of symmetry in a form that would be recognised today was presented by the British mathematician Arthur Cayley in the middle years of the nineteenth century.

Mathematicians understand symmetry in terms of transformations, such as a reflection in a mirror, a rotation around an axis or a translation – a smooth shift – through space. An object or a pattern possesses a symmetry if it remains completely unchanged after it is transformed in some way. In the words of the mathematician Hermann Weyl, 'A thing is symmetrical if there is something we can do to it so that after we have done it, it looks the same as it did before.'[34] For instance, if a snowflake is rotated around its centre by

60 degrees, or ⅙ of a full rotation, it will appear exactly the same after the rotation as it did before the rotation. We would not be able to see any difference in the snowflake before and after the transformation. This is what mathematicians mean when they say that the snowflake is symmetrical or invariant under a 60-degree rotation, or that the 60-degree rotation is a symmetry of the snowflake. Of course, the snowflake is symmetrical under other transformations as well – it is also symmetrical under a rotation of 120 degrees, or ⅓ of a complete rotation. In fact, for a snowflake, a rotation through any multiple of 60 degrees is a symmetry. The snowflake also has reflection symmetries. If the snowflake were reflected in a hypothetical mirror positioned on a line joining opposite vertices, the two halves of the snowflake would swap over. This would not alter the appearance of the snowflake, so it is a reflection symmetry of the snowflake.

The full collection of all the symmetries of an object is called its symmetry group, and the area of mathematics devoted to the study of symmetry is known as group theory.[35] Group theory is one of the most elegant and important branches of mathematics. It is founded on a set of fundamental statements, its axioms, in the style of the classical Greek geometers as championed by Euclid. Just like classical geometry, group theory is a beautiful edifice that mathematicians have built up brick by brick. Pure mathematicians study it for its own intrinsic allure, but it has also developed into a very powerful mathematical tool that is indispensable to the modern physicist.

Multiple reflections in the two mirrors of Brewster's original kaleidoscope give rise to the full group of symmetries of the hexagon, which is why the images produced by the kaleidoscope all have this hexagonal symmetry. Adding further mirrors that are positioned appropriately will produce all the reflections of other, larger symmetry groups. For example, a third mirror in the correct position will produce all the reflection symmetries of a tessellation, a pattern that fits together like a regular array of tiles and covers a flat surface, as shown in Plate 1 (lower half). The reflection group of the hexagon is contained within this larger symmetry group. Groups typically fall into nested hierarchies in this way.

With three triangular mirrors we can produce a kaleidoscope whose multiple reflections will generate all the symmetries of a regular three-dimensional object. A computer-generated example of such a kaleidoscopic image is shown in Plate 2. The pattern on the sphere has the same symmetry as two of the Platonic solids, the icosahedron and the dodecahedron. The three virtual mirrors that were used to make the kaleidoscope were positioned to form three of the four faces of a tetrahedron,[36] with the fourth face left open. The image was produced by placing part of a multicoloured sphere within the mirrored tetrahedron and then viewing the resulting pattern of reflections through the open face of the tetrahedron.

If a fourth mirror is added to complete the virtual tetrahedron, it will generate the symmetries of a pattern that fills the whole of three-dimensional space, as shown in Plate 3. This image shows the view from within the virtual tetrahedron. A single blue cylinder has undergone multiple reflections in the four mirrors to produce images of all the other cylinders. These cylinders are the edges of two types of polyhedra, octahedra and cuboctahedra, that fit together to form a space-filling structure known to mathematicians as a honeycomb.[37] The image gives the impression that we are located inside the honeycomb, which recedes to infinity in all directions. This honeycomb is closely related to Kepler's packing of spheres and therefore also to the atomic structure of many crystals.[38]

Most of the groups studied by mathematicians describe symmetries that are not so readily depicted. The abstract world of mathematics does not stop with just four mirrors. Typically, mathematicians analyse groups that encapsulate the symmetries of higher-dimensional objects that cannot be constructed in three-dimensional space. The power of mathematics is that it enables such abstract entities as these to be studied, even though their full multi-dimensional symmetry cannot be directly visualised. It is quite remarkable that physicists have found that some of these abstract higher-dimensional groups are very useful when it comes to describing the structure of the real world.

Symmetry has turned out to be fundamental to the way physicists view the universe, and a better understanding of the symmetries of nature has been one of the predominant themes in the development of physics. When the structure of matter is analysed, symmetry and patterns become evident at all levels. These patterns may arise because the components of matter are arranged in a regular way, which means that the presence of symmetry often provides physicists with a clue to the existence of a deeper level of structure. Kepler was the first to realise this when he suggested that crystals have a regular, geometric appearance because they are composed of regular, geometric arrays of atoms. We will meet further examples of symmetry in later chapters when we delve deeper into the structure of matter. We will also see that the forces that hold the components of matter together can also be understood in terms of abstract symmetries.[39]

Symmetry and Conservation

In the medieval universe, the Earth was a special place: it sat motionless at the centre while the heavens revolved around it. Modern physics is much more democratic. All points in the universe are considered to be equal in the sense that we

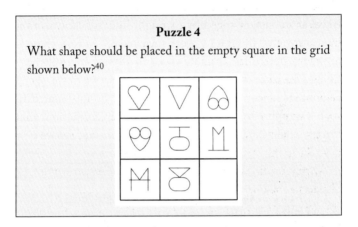

Puzzle 4

What shape should be placed in the empty square in the grid shown below?[40]

expect the fundamental laws of the universe to apply in the same way everywhere. We must therefore formulate these laws in such a way that they are symmetrical under translations through space – which is merely common sense, as nobody would expect the results of an experiment to be different if it is carried out first in Europe and then several thousand kilometres away in the United States. It is also natural to expect that the laws of physics must remain the same from moment to moment. Nobody would expect the result of an experiment in fundamental physics to depend on the day on which the experiment was performed.

Just over a hundred years ago, it was Einstein who first recognised the importance of explicitly formulating physics in a way that makes such symmetries apparent. At the heart of Einstein's theory of special relativity is the idea that physics appears the same to observers who are moving at different speeds. We notice when our speed in any direction changes: we may feel a sharp acceleration in a sports car, or the jostling of our train as it passes over a set of points, or the buffeting of our aircraft as it hits a pocket of turbulence. But as long as we travel at a constant speed in a straight line, we can function in every way as though we were stationary.[41] In fact, there is no way to determine absolute speed in physics: it is only possible to measure the speed of one object relative to the speed of another object. Physics must therefore be formulated in such a way that it is symmetrical under transformations that change one constant velocity to another constant velocity.

Answer to Puzzle 4

The shapes in the grid are each of the digits from 1 to 9 combined with their mirror image and arranged in a 3×3 magic square. The missing digit is 8, so the missing symbol is 8̸.

At first glance, these examples of symmetry seem fairly obvious, and it is not clear that they provide any significant insight into physics at all. However, the apparently innocuous statement that physics is the same at all points and all times actually places very powerful constraints on the way the universe works. The strength of the statement is apparent from a remarkable result discovered by a female mathematician called Emmy Noether (pronounced 'nurter'). Of course, the fact that Noether was female should not have mattered at all, but unfortunately it did because she had to struggle for recognition throughout her life simply because of her gender. Even with the support of some of Germany's leading mathematicians, the conservative academic tradition in her homeland a century ago placed serious obstacles in her way. Despite this, in 1915, while studying Einstein's brand-new general theory of relativity, she proved her profound result, which has since become a cornerstone of modern physics.

What Noether discovered was a fundamental link between symmetry and conservation laws, such as the law of conservation of energy. This connection means that for each symmetry there is a corresponding quantity that must remain constant. For instance, the invariance of the laws of physics under changes in position implies that the total momentum of a collection of interacting bodies cannot change; similarly, the requirement that the laws are the same today as they will be tomorrow or at any other time means that energy must be conserved in all processes. This is quite an incredible statement. We know that in any conceivable process, no matter how complicated, if we add up all the energy beforehand, then the total energy afterwards will be exactly the same. Energy can neither be created nor destroyed; it can only be converted from one form into another. According to Noether, the reason that we know this to be true is simply that we expect the fundamental laws of the universe to be the same at all times. (Hopefully this won't come as too much of a disappointment to anyone who had been planning to build a perpetual-motion machine to solve the world's energy problems.)

Conservation laws are among the most valuable tools that physicists have. This is why arguments based on symmetry are so powerful. When a quantity remains unchanged throughout a physical process, it helps physicists to disentangle the details of what might be an extremely complicated event. This is especially true of the interactions that take place in a particle accelerator such as the LHC. For example, electric charge is conserved in all physical processes. Take a collection of particles, add up the electric charges on each particle, then allow the particles to interact with one another. In the interaction some particles might be destroyed and others created, but the sum of the electric charges at the end of the interaction will be exactly the same as it was before. Whether the process is a chemical reaction in a test tube, a nuclear reaction in the Sun or a collision in a particle accelerator, the total electric charge will always be conserved.[42]

In the middle years of the twentieth century, two new forces were discovered that are responsible for radioactivity and the physics of the atomic nucleus. These two forces are known as the weak force and the strong force. Particles such as electrons and quarks carry charges corresponding to each force, and these charges operate in very much the same way as electric charge. They determine how the particle will respond to each force, and they are conserved when particles interact. As I will reveal in later chapters, the catalogue of fundamental particles known to physicists is quite small. Each particle possesses a number of properties that distinguish it from other particles, and those properties are simply all the different charges that the particle carries. So these lists of charges can be used to identify a particle. When our two protons blasted themselves apart in their head-on collision at the start of this chapter, the result was a complicated melange of new particles racing away from the impact point. To track down the Higgs within the debris of this explosion, the physicists at CERN must use the conservation laws to identify each particle. In fact, this procedure is now routine; there are so many particles spraying through the detectors at the LHC that the task is delegated to the electronics within the detectors, which can accomplish it within a nanosecond of each collision.

Einstein believed that the ultimate laws of nature must be as symmetrical as possible, with the widest range of natural processes explained in the most economical way.[43] He said: 'It is a wonderful feeling to recognize the unity of a complex of phenomena which appear to direct sense observation as totally separate things.'[44] Einstein's later years were spent in a quest to completely unify all the forces of nature. He was unsuccessful in this incredibly ambitious project, but his philosophy remains at the heart of modern theoretical physics. It continues to set the agenda for the challenge of unifying the forces and constructing the most symmetrical and perfect account of their operation. The Higgs particle is a leading player in this drama. Its role is to unify two of the three forces that operate in particle interactions. The discovery of the Higgs confirms the belief of physicists that the electromagnetic force and the weak force are really just two manifestations of a united electroweak force, and this represents a major step towards the ultimate goal of unifying all the forces. The next chapter explains why unification matters and describes the dramatic impact that its previous successes have had on all our lives.

Chapter Two

UNITY IS STRENGTH

In the year 1666 he retired again from Cambridge …
[He went] to his mother in Lincolnshire and whilst he
was musing in a garden it came into his thought that the
power of gravity (which brought the apple from the tree
to the ground) was not limited to a certain distance from
the earth but that this power must extend much farther
than was usually thought. Why not as high as the moon
said he to himself and if so that must influence her motion
and perhaps retain her in her orbit, whereupon he fell a
calculating what would be the effect of that supposition.

John Conduitt, *Account of Newton's Life at Cambridge*
(*c.*1727/28)[1]

It Was Gravity that Pulled Us Down

Today, we take it for granted that the force that pulls an
apple to the ground when it ripens and falls from a tree is

the same as the force that holds the Moon in its orbit around the Earth. This force is, of course, gravity. But the unity of these two forces was not obvious to the great minds of ancient Greece. Indeed, it was only three and a half centuries ago that Isaac Newton conclusively demonstrated that these two forces are identical. In his old age, Newton told the story that he had been inspired to consider these matters as a twenty-three-year-old student sitting beneath an apple tree in the village of Woolsthorpe in the summer of 1666. Several authors recorded the story including John Conduitt, the husband of Newton's niece Catherine, who wrote the extract above. But we know from Newton's notes that he had been thinking constantly about mechanics and mathematics for the previous two years, so his understanding of gravity did not really come as a sudden flash of inspiration under a tree. Newton presented to the world the final distillation of his theories of motion and gravitation when he published his *Philosophiae Naturalis Principiae Mathematica* ('Mathematical Principles of Natural Philosophy') in 1687. The *Principia*, as it is usually known, was a revelation. Physics would never be the same again. Newton's theories became the model for how the exact sciences, with laws amenable to accurate quantitative expression, should be formulated.

In the *Principia*, Newton proposed that the force of gravity operates in the same way on all objects. It is universal and applies to all bodies everywhere in the universe. Newton realised that this elegant idea would enable him to explain a diverse range of very different observations as the result of the action of just one force, the force of gravity. He could account for the motion of projectiles near the Earth's surface as well as the motion of the planets in their orbits around the Sun. He could calculate the courses of comets across the sky. He could also explain the origin of the tides as the result of the gravitational attraction of the Moon and the Sun on the Earth's oceans. Newton's insight offers us a unified and therefore greatly simplified view of these forces and their role in the universe. A diverse range of phenomena are all explained by the operation of a single gravitational force.

The dramatic transformation in our knowledge of the fundamental workings of the universe since Newton's time is derived from our improved understanding of the forces of nature. In particular, several significant breakthroughs have followed from the realisation that forces previously considered to be distinct are really just different manifestations of the same fundamental force. Each example of the so-called unification of forces has its own individual character. A particularly illuminating one is the unification of the electric and magnetic forces. The recognition that all electric and magnetic forces are simply different faces of a single electromagnetic force arrived with the theories of Michael Faraday and James Clerk Maxwell in the middle of the nineteenth century. The direct result of this great theoretical advance was a technological revolution. It is hard to imagine how we could function today without a ready supply of electricity. But without a firm understanding of the interplay between electricity and magnetism, as revealed by these nineteenth-century pioneers, we would have no refrigerators, televisions, computers, mobile phones, cappuccino machines or all the other electrical devices that seem indispensible to modern life.

These great Victorian scientists were following in the footsteps of many earlier researchers. Magnetism and static electricity were both known to the ancient Greeks. The word 'magnet' is derived from the name of the iron ore 'magnetite', which was notable for its magnetic properties. This mineral got its name from the Greek region of Magnesia in Thessaly, where it was mined. The word 'electricity' also has a Greek origin. If a piece of amber is rubbed with a cloth, both the cloth and the amber become electrically charged. We are familiar with this 'static electricity', as it is called. When we brush our hair, static electricity can make some of the hair stand up; if we rub our shoes across a thick carpet we may become electrically charged and receive a small shock when we touch a metallic object and the charge is dissipated. The Greek word for amber is *elektron*, and this is the origin of the word 'electricity'.

A Magnetic Character

The scientific investigation of magnetism dates back even further than Newton's time, to the Tudor physician William Gilbert, who was born in Colchester in 1544.[2] He received his university education at St John's College, Cambridge, and following a period as college bursar he left to practise medicine in London, where he was elected to the presidency of the College of Physicians in 1600. From 1601 until his death in 1603 he was the personal physician of the monarch. He was first appointed by Elizabeth I, and his services were retained by James I on his accession to the English throne in 1603.

Gilbert was one of the first scholars to attempt to understand the workings of nature by experimentation. He passionately rejected the traditional Aristotelian philosophy and the established method of university teaching that relied on philosophical argumentation rather than scientific investigation. Gilbert presented his ideas about magnetism in a book published in 1600. Its full title is *De Magnete, Magneticisque Corporibus, et de Magno Magnete Tellure* ('On the Magnet and Magnetic Bodies, and on the Great Magnet the Earth'),

Figure 11 A modern representation of William Gilbert's *terrella* as a magnetised model Earth.

but it is known as *De Magnete*. In this short book Gilbert describes how he made a model Earth, which he called a *terrella*, or 'little Earth', out of a magnetised ball of iron. He placed magnetised needles around the *terrella* and discovered that they always pointed towards the poles of this magnetised globe (see Figure 11). Gilbert is believed to have demonstrated this marvel to the curious Queen Elizabeth.

Gilbert's most important conclusion from his many experiments with the *terrella* was that the Earth itself is an enormous magnet, and that this explains why compass needles point northwards. His idea was completely novel, but it was also correct. None of Gilbert's contemporaries could offer a plausible alternative explanation for this property of a magnetised needle. Some attributed the behaviour of compass needles to an astrological attraction to the pole star; others proposed the fanciful idea that there might be a large magnetic island at the North Pole. Although Gilbert played a very important role in putting electric and magnetic forces on a firm scientific footing, he believed that they were quite different and unrelated. It would be another 250 years before the relationship between the two forces was understood. On 30 November 1603, three years after the publication of *De Magnete*, Gilbert died, possibly in one of London's recurrent outbreaks of bubonic plague.

The Electric Force

It took the efforts of many great researchers over the course of several centuries to uncover the secrets of electromagnetism. Maxwell finally expressed the united nature of the electric and magnetic forces in a mathematical theory that encapsulated experimental results from many years of investigation by the experimental physicist Michael Faraday. This synthesis of the electric and magnetic forces is the classic model for the unification of forces and remains the inspiration for modern attempts to further unify the forces.

Before delving into the story of their unification, here is a quick outline of how the electric and magnetic forces operate. All bodies feel the effect of gravitation, because it is the mass of the body that is the source of the gravitational force. The electric force is different: only bodies that are electrically charged feel the electric force. An uncharged body will not be subject to any electric force. At a fundamental level, the electric charges are carried by the elementary particles, such as electrons and protons, from which matter is composed. There are two types of electric charge, which we call positive and negative charge. For instance, the charge on an electron is negative and the charge on a proton is positive. The electric force between two particles depends on the electric charges of both particles. If the two particles have the same type of charge, the force between the particles will be repulsive. If the two particles have different types of charge, then the electric force between the two particles will be attractive. So two electrons will repel each other, and two protons will repel each other, but an electron and a proton will attract each other.

The electric force is well known to us as the source of electricity. An electric current is a flow of electrically charged particles, and these particles are usually electrons. We have an advantage over Faraday and Maxwell and the other early investigators of electricity because we know about the fundamental particles that carry the electric charge, so the origin of electricity is much clearer to us than it was to them. The electron was not discovered until 1897, several decades after Maxwell had published his theory of electromagnetism.[3]

The magnetic force is most familiar to us from playing with bar magnets. A bar magnet has a north pole and a south pole.[4] If we hold one bar magnet and move another bar magnet towards it, we can feel the two north poles repelling each other and the two south poles repelling each other, and with one of the magnets inverted we can feel the north pole of one magnet attracting the south pole of the other. The magnetic force is similar to the electric force in that, just like electric charges, magnetic poles of the same type repel, but poles of opposite type attract. One big difference between the magnetic force and the electric force is that the two poles of the magnet cannot be separated. If you cut a bar magnet in half, the two halves will still have both north and south poles.

The Voltaic Pile

It was the eighteenth-century American scientist and states-man Benjamin Franklin who first convincingly demon-strated the connection between lightning and electricity.[5] He flew kites in thunderstorms to investigate the origin of lightning and was able to prove that lightning was an elec-trical effect, identical in nature to the electrical discharges produced by mechanical devices. Luckily for Franklin and the American nation, he also managed to avoid killing himself in the process. He exploited his findings by inventing the lightning conductor.

The study of the electric force gathered pace towards the end of the eighteenth century when an Italian Count, Alessandro Volta, discovered a means of producing a control-lable supply of electricity. This enabled researchers to conduct systematic electrical experiments that would prove that the various apparently different types of electricity were actually identical. Volta's investigations had been sparked by a discov-ery made by the Italian anatomist Luigi Galvani (1737–98) while he was dissecting frogs. Galvani demonstrated his dis-covery with an implement he had constructed from a zinc rod and a copper rod joined together to form a pair of calli-pers or dividers. He found that if he simultaneously touched the nerve in a frog's dissected leg with the zinc rod and the frog's leg muscle with the copper rod, the muscle would con-tract, just as if the frog were alive. This effect was described as 'animal electricity' by Galvani. He believed that it was a purely biological effect produced by an electrical fluid that flowed through the nerves of living organisms. He saw it as a feature of a biological life force that animated inert matter.

Volta was intrigued by Galvani's discovery, but he didn't accept his explanation. Volta believed that the flow of electricity was simply a physical effect, a consequence of the properties of the two metals from which Galvani had constructed his callipers, and in order to prove this, he constructed a device called a 'pile'. Volta's pile was a stack of discs composed of two different metals. The original voltaic pile consisted of alternating zinc and copper discs, with each disc separated from the next by a layer of paper moistened with salt water. When the top and bottom discs were connected by a conducting wire, an electric current would flow through the wire. Volta had demonstrated that the electrical effects observed by Galvani were simply the result of the construction of the callipers from two different types of metal. It was a physical effect produced by the chemistry of the metals, and not a specifically biological effect. Although Galvani was correct in his observation that electricity plays a role in the functioning of nerves and muscles, his 'animal electricity' was no different to the electricity produced by Volta's pile.

The modern term for the voltaic pile is a battery. Volta's invention was hugely important because scientists now had a supply of electricity whenever they needed it, and could now embark on a thorough investigation of electricity in their laboratories. After Volta's conclusive demonstration of the equivalence of 'animal electricity' and the current from his battery, other investigators demonstrated further connections. For instance, they showed that the current from one of Volta's batteries would produce the same effects as static electricity, generated by rubbing amber or with some type of mechanical device. If someone held the two ends of a wire from one of Volta's batteries, they would feel a tingling sensation in their fingers and could be given small electric shocks. The current could also be used to produce a spark, like lightning, but on a very small scale. So by the early years of the nineteenth century, researchers were establishing the identity of the different types of electricity. The conversion of electricity from one form to another showed that what had seemed to be different types of electricity were actually the same. The electrical force was becoming recognised as a single, unified force.

The capacity of Volta's pile to produce electricity through the chemistry of its components had deep implications. It suggested that the electric force plays a fundamental role in the structure of matter. A young English scientist took advantage of these early batteries and their potential to offer electricity on demand and pioneered the use of electricity to investigate the chemical foundations of matter. This was Humphry Davy.

Humphry Davy

> Sir Humphry Davy
> Abominated gravy.
> He lived in the odium
> Of having discovered sodium.
>
> Edmund Clerihew Bentley (1905)[6]

Humphry Davy would go on to hold a highly prestigious professorship in chemistry at the Royal Institution, but his start in life was modest. He was born in 1778 in Penzance, Cornwall, the son of a penniless woodcarver. After completing his schooling in Penzance and Truro, he was apprenticed to a local apothecary. The young Davy was introduced to a Dr Thomas Beddoes, a philanthropic physician based in Bristol, who was on a geological pilgrimage to the rocky coast of Cornwall. Beddoes must have been very impressed by the precocious youth, because in 1798, when he founded the Pneumatic Institute in Bristol, he hired Davy as his assistant to help him investigate the possible therapeutic effects of various newly discovered gases, then called 'factitious airs' or 'vapours'.

Davy moved into Beddoes's home in the fashionable neighbourhood of Clifton. The house was equipped with an excellent laboratory where Davy could conduct chemical experiments. Beddoes would administer a variety of gases to his patients with the aim of curing diseases such as consumption. Davy worked in the laboratory and was given the hazardous task of determining the effects of inhaling different gases and classifying them as either stimulants or sedatives. Beddoes had many eminent and wealthy patients. Among the numerous characters to pay social visits to his home were the Romantic poets Robert Southey and Samuel Taylor Coleridge. Both became close friends of Davy.

One of the gases Davy examined was the newly discovered nitrous oxide, first isolated by Joseph Priestley. It was a colourless, non-flammable gas with a slightly sweet smell. Davy discovered that the gas had remarkable properties. When inhaled it immediately induced a feeling of euphoria, an effect which gave it its common name, 'laughing gas'. On one occasion when Davy was suffering with toothache, he noticed that the gas was a powerful painkiller: 'As nitrous oxide in its extensive operation appears capable of destroying physical pain, it may probably be used with advantage during surgical operations in which no great effusion of blood takes place.'[7]

Although Davy's writings include this suggestion for the medicinal use of nitrous oxide, it was not taken up as an anaesthetic for many years, but eventually it did become the standard painkiller used throughout the world for dental surgery. Davy demonstrated the effects of nitrous oxide at social gatherings, and soon he, Southey, Coleridge and other acquaintances were using it for recreational purposes. Davy eventually became addicted to its intoxicating effects and continued to use it throughout his life. He claimed that it had all the benefits of alcohol, but with none of the drawbacks.[8]

By the age of just twenty-one, Davy was well known in scientific and social circles. He soon caught the attention of an American scientist with a Bavarian title and a British knighthood, Sir Benjamin Thompson, also known as Count Rumford. Thompson would later marry Marie-Anne, the widow of the French chemist Lavoisier and a great chemist in her own right. In 1800 the enterprising Thompson helped to found the Royal Institution in London with the aim of stimulating scientific discoveries and promoting their industrial application. The Institution was provided with a well-equipped laboratory and a lecture theatre, and Humphry Davy was appointed assistant lecturer in chemistry. Within a couple of years he was promoted to professor of chemistry at the Royal Institution, and in 1804 he was elected to a Fellow of the Royal Society.

From the very beginning, one of the Royal Institution's aims was to promote science to the public, and Davy was soon much in demand as a lecturer. His charismatic style of presentation completely captivated his audiences and he soon gained a reputation as a skilful orator. His talks were attended by wealthy members of London society who were willing to pay large sums for their tickets. The glamorous Davy became so popular that when he fell ill his concerned female admirers created chaos with their horses and carriages blocking the streets around the Royal Institution. To meet the demand for news about his condition, the staff set up a board outside the building on which were posted medical reports, temperature charts and up-to-date accounts of his progress.[9]

As director of the Royal Institution, Davy took up the fashionable study of electricity using the newly invented voltaic piles. This research led to his greatest achievements: the discovery of a number of new chemical elements. Among the list of chemical elements drawn up by Lavoisier were several that he referred to as 'earths'. Davy surmised that these earths might not in fact be true elements, and that it might be possible to decompose them with electricity. He took one of the earths, known as potash, dissolved it in water and passed electricity through the solution. Oxygen and hydrogen were given off. He next tried passing electricity directly through solid potash. He took some fresh potash and placed it on a platinum disc that was connected to the negative terminal of his battery. He then connected the positive terminal of the battery to a platinum wire and plunged the wire into the potash so that an electric current ran through it. Within a few minutes, little metallic globules were appearing on the platinum disc. At this point, according to his assistant, Davy did a jig: 'He bound around the room in ecstatic delight, and some little time was required for him to compose himself sufficiently to continue the experiment.'[10] Davy would christen the metal 'potassium'. This was the first time a new metal had been isolated by electrolysis – the electrical decomposition of a chemical substance. To ensure that the metal did not originate in the platinum electrodes, Davy repeated the experiment using different metals as his electrodes. The result was always the same. He had discovered a new element.

In the same month, October 1807, Davy passed an electric current through caustic soda, another one of Lavoisier's earths. Davy recorded that it was necessary to use a higher electric current to decompose the caustic soda than was required in his experiment with potash. Again, the outcome was the production of globules of a new metallic element, 'sodium'. The following year, Davy used the same technique to extract yet another new metallic element, calcium.[11] He had demonstrated beyond all doubt that the electric force plays an important role in chemical reactions and the fundamental structure of matter.

Davy was knighted by George III in 1812 and made a baronet in 1818. He was elected president of the Royal Society, a role in which he served from 1820 to 1827. But despite his extremely impressive career and his many outstanding achievements, Davy would be overshadowed in the history of science by his protégé, Michael Faraday.

Michael Faraday

Michael Faraday was born in 1791 in one of the poorer districts of London, south of the Thames.[12] He was the son of James Faraday, a journeyman blacksmith originally from the village of Clapham in Yorkshire. Faraday's early years were extremely hard. Often the family had to subsist for a week on nothing more than a single loaf of bread each. The education that Faraday received was consequently very rudimentary. But in 1804, when he was thirteen, he had a stroke of good fortune: a local bookseller hired him as an errand boy, and the following year he was indentured as the bookseller's apprentice to learn the trade of a bookbinder and stationer.

All the young Faraday's spare time was spent reading avidly in the bookshop, where he devoured books on all topics in an effort to make up for his poor formal education. Among the many books stocked in the shop were some of the latest works of scientific literature, and Faraday was particularly impressed by the chemical philosophy of Humphry Davy. Inspired by this literature, Faraday began to attend public lectures on science whenever his limited funds would allow it, and in the evenings he started to conduct simple experiments in a spare room in the shop. Throughout his life, he would keep accurate records of his experiments. The first surviving account of such an experiment is in a letter to a friend written in July 1812 and describes the construction of a voltaic pile.[13] The pile was formed from seven copper halfpenny pieces, stacked alternately with discs of zinc and interleaved by pieces of paper moistened with brine. Faraday passed the electric current generated by his pile through a solution of magnesium sulphate and succeeded in decomposing this chemical compound.

One fateful day in 1812, a visitor to the bookshop who was aware of Faraday's scientific interests offered him tickets to a series of lectures by his great hero, the flamboyant professor of chemistry Sir Humphry Davy. Faraday gratefully accepted and attended the lectures, at which he took meticulous and extensive notes. Harnessing his skills as a bookbinder's apprentice, Faraday bound the three hundred pages of lecture notes that he had compiled and sent them off to Davy. The great professor must have been very impressed, as Faraday received an immediate reply and an offer of an interview. Shortly afterwards Davy's eyesight was damaged in an explosion in his laboratory, and he decided to employ Faraday as his secretary.

Before long, Faraday obtained the position of scientific assistant to Davy and began to work alongside him in the laboratory. In October 1813, Davy and his wife embarked on the Grand Tour, and they took Faraday along as their companion. Faraday would later complain that he was treated more as the valet rather than Davy's scientific companion, but he was grateful for the opportunity to learn about Davy's scientific philosophy and his approach to experimental investigation. On their way around the great European cities, Davy and Faraday met many of the leading scientists of the day, including the Count Alessandro Volta and another pioneer in the study of electricity, André-Marie Ampère.

Electric and Magnetic Fields

Although Faraday's mathematical knowledge was limited because he had not received a good formal education, it still seems incredible that there isn't a single mathematical formula to be found in any of Faraday's notebooks.[14] What he did excel at was representing his ideas pictorially and expressing them in clear and simple language. One outstanding contribution to science was his invention of the concept of a field, which he used to describe the forces he explored in his experiments. He imagined that an electrically charged body is surrounded by an electric field that permeates the whole of space. Faraday visualised this field as a map of the electric forces around the charged body. He drew maps indicating the direction in which a force would be felt at different points in space by a positively charged particle that might be situated at each of those points. The electric field is therefore a hypothetical map that shows the force at every point in space. Despite being much criticised when he introduced it, Faraday's field concept has become an integral part of our modern understanding of all physical forces.

As an example, we can draw a map of the electric force around a negatively charged particle, such as an electron. The map is a collection of lines, each with an arrow on it, that show the direction in which a force would be felt at every point around the particle. To construct our map, imagine placing a positively charged 'test' particle at each point in space, then plotting the direction of the electric force felt by the test particle. We can use this information to draw the map – the electric field – of the negatively charged particle. Although there is a line of force through every point in space, the map will just show a representative sample of these lines. We can easily work out what the map will look like. The force on the positive test charge will be directed towards the negatively charged particle, so the map will show the negatively charged particle situated at the centre of a collection of lines pointing radially inwards, as shown on the left in Figure 12. This is the electric field of the negatively charged particle. The diagram actually shows a two-dimensional cross section of the electric field; in reality the field lines extend outwards in all directions in three-dimensional space.

What would the map of the electric field around a positively charged particle look like? The force between the test particle and another positively charged particle is repulsive. The field is again radial, but all the arrows are now pointing outwards, as

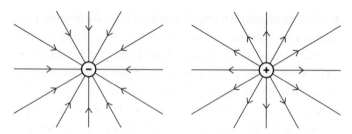

Figure 12 Left: The electric field around a negatively charged particle. Right: The electric field around a positively charged particle. (In this diagram and the ones that follow, what is shown is a two-dimensional cross section through a three-dimensional field.)

this is the direction of the force that would be felt by the positively charged test particle, as shown on the right in Figure 12.

Faraday's maps also contain information about the strength of the force. This is indicated by the density of the lines in any region of the map. The force is strongest close to the charged particle. We can see this in our map, because the lines crowd closer together as they approach the particle.

What about the electric field around two charged particles, one positively charged and one negatively charged? We can map out the field as before by imagining the force that would be felt by a hypothetical positively charged test particle situated at each point in space. The total force on the test particle will now be a combination of a repulsive force away from the positively charged particle and an attractive force towards the negatively charged particle. The result is shown in Figure 13. Close to the positively charged particle, the field lines emanate radially outwards from the positive charge; close to the negatively charged particle, the field lines are directed radially inwards towards the negative charge. Farther from the two particles, the direction of the force is determined by the relative strengths of the force towards the negative charge and the force away from the positive charge.

Maps of magnetic fields can be made in just the same way. Faraday drew maps of the magnetic field lines around a bar magnet which looked like Figure 14. The bar magnet has a north pole and a south pole. Although the north and south poles of the magnet cannot be separated, we can consider

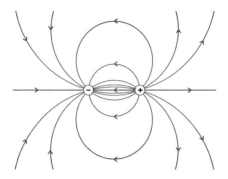

Figure 13 The electric field around a pair of charged particles, one positively charged and the other negatively charged.

them to be the magnetic equivalents of the positively charged particle and the negatively charged particle in the previous figures. The magnetic field lines around the bar magnet turn out to have the same shape as the electric field lines around the pair of oppositely charged particles shown in Figure 13. Of course, the field lines now indicate the direction of the magnetic force, not the electric force.

Electromagnetism

At the time when Faraday was working as Davy's assistant, electricity was becoming increasingly well understood. The leading researchers recognised that although electricity might appear in different guises, ultimately these were all manifestations of a single force. But although the electric force was unified at around this time, the relationship between electricity and magnetism remained a mystery. No one knew whether magnetism bore any relationship to the electric force at all, and although similarities between the two forces could be

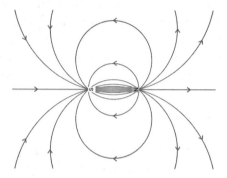

Figure 14 The magnetic field around a bar magnet.

pointed to, it was clear that magnetism was not simply another form of electricity. There were a number of scientists, however, who believed in the fundamental unity of the laws of nature and who were convinced that a link between electricity and magnetism must exist. One of these was the Danish physicist Hans Christian Ørsted, and he was the first to notice an effect that clearly demonstrated just such a link.

One evening in April 1820, Ørsted was preparing to give a public lecture when he observed a surprising phenomenon. The experiment he was setting up was designed to demonstrate some of the properties of electricity, so the apparatus on his desktop included a wire circuit and a battery that would supply the electric current. Also on the desktop was a magnetic compass. As he was testing his equipment before the lecture, he noticed that whenever he closed the switch in the circuit and an electric current flowed through the wire, the compass needle moved to point in a different direction. When he switched the current off, the needle returned to its original north–south alignment. He had established an unmistakable connection between electricity and magnetism. Ørsted published his findings and triggered a flurry of excited activity in laboratories throughout Europe. All over the continent, researchers took up the investigation of this new discovery, seeking to work out the implications for the laws of nature of a link between electricity and magnetism.

The Homopolar Motor

The following year, Faraday was asked to prepare a 'Historical sketch of electromagnetism' for the journal *Annals of Philosophy*.[15] In preparation for writing this review article, the diligent Faraday set out to reproduce all the experiments on the subject whose results had previously been published.[16] The most important were those reported by Ampère, who had become captivated by the study of electricity and magnetism since Ørsted's recent breakthrough. Ampère had observed that a wire carrying a current produces a force on a magnet that will try to make the magnet rotate around the wire. Alternatively, if the magnet is fixed, then the force will move the wire carrying the current in a circle around the magnet.

In September 1821, Faraday devised an elegant experiment that beautifully confirmed Ampère's findings. Electricity from a voltaic pile was passed through a beaker of mercury. This was the clever part of the experiment because mercury is the only metal that is liquid at room temperature, and because it is metallic it conducts electricity. The mercury thus formed part of Faraday's electric circuit, and when the circuit was closed, the current flowed through the mercury. To complete the circuit, he directed one wire from the pile through a hole in the base of the beaker so that it was in contact with the bottom of the mercury, and he fixed the other wire from the pile to a cork floating in the beaker so that this wire was in contact with the surface of the mercury. A bar magnet was then placed in the beaker with its bottom end fixed and its top end free to move, as shown in Figure 15. When the current from the pile or battery was passed through the circuit, it produced a force on the magnet causing its free end to move through the mercury in a circular path. If the wires were swapped over so that the current flowed in the opposite direction, the magnet would rotate in the opposite direction.[17]

Figure 15 Faraday's rotating magnet experiment: the first homopolar motor.

Faraday had constructed the first ever electric motor – a device that is called a homopolar motor. Faraday could visualise the magnetic field generated by the electric current that was driving the homopolar motor. The magnetic field lines form circles that are perpendicular to the direction of flow of the electric current, as shown in Figure 16. The north pole of the magnet feels a magnetic force that causes it to rotate in the beaker.

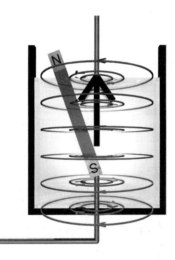

Figure 16 Homopolar motor. The large arrow indicates the direction of flow of the electric current; the magnetic field lines form circles around the flowing current.

Faraday's experiment gives a clear picture of the magnetic field around an electric current. We can draw a similar map of the magnetic field in the more familiar case of an electric current flowing through a wire. The north pole of a magnet placed close to the wire carrying the current feels a force that will tend to move it around the wire, as Ampère had

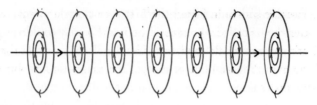

Figure 17 The magnetic field around a wire carrying an electric current.

observed in his experiments. And this is what is indicated in the map of the magnetic field, as shown in Figure 17.

Faraday's diagrams now enabled him to see a connection between the magnetic field produced by a current in a wire and the magnetic field of a bar magnet. If a wire carrying an electric current were bent into a loop, the magnetic field lines would be compressed within the loop and the field would look like Figure 18. This magnetic field looks similar to the one around a bar magnet, shown in Figure 14. Faraday now realised that by coiling the wire around so that the electric current passes through many loops, it would be possible to produce a much stronger magnetic field that would be indistinguishable from the magnetic field of a bar

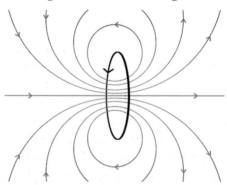

Figure 18 The magnetic field around a single loop of wire carrying an electric current.

magnet, as shown in Figure 19.[18] This current-carrying coil is known as an electromagnet. The construction of a strong electromagnet would lead Faraday to another great break-through in August 1831, possibly the most important of his many discoveries.

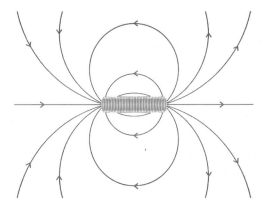

Figure 19 The magnetic field around a copper coil carrying an electric current.

Induction

Faraday was a firm believer in the underlying symmetry of nature. He conjectured that since an electric current produces a magnetic field, it must also be possible to use a magnetic field to generate an electric current.[19] To test this hypothesis, he built a powerful electromagnet by winding an insulated wire around a thick iron ring. He then wound a second coil of wire around the first and attached it to a meter to measure any current that might flow through it. Faraday's next step was to pass a current through the primary coil of wire to create a magnetic field within the ring. He expected the magnetic field generated by the current flowing in the primary coil to induce a continuous current in the secondary coil, but that was not what he found. Faraday discovered that the needle of the meter connected to the secondary coil was indeed displaced, but only at the moment when the current in the primary coil was switched either on or off. This was a big surprise.

Faraday interpreted his findings in the following way: when the switch was turned off, the current in the primary coil quickly fell to zero. This rapidly changing electric field generated a brief pulse of magnetic field. The sudden appearance and disappearance of this magnetic field generated an equally brief electric field, and this electric field produced a short pulse of current in the secondary wire. His crucial observation was that although a steady current will generate a magnetic field, a changing magnetic field is necessary to generate an electric field. Faraday had discovered electromagnetic induction. The coil he made for this experiment can still be seen on display at the Royal Institution in London. This apparatus is the prototype of the electrical transformer, a piece of equipment that would be vital for the development of the future electrical industry.

Guided by his new insight that a changing magnetic field will generate an electric field, Faraday's progress was both rapid and very systematic. From his meticulous notebooks we know exactly when he carried out each experiment and how his ideas developed. On 17 October 1831 he generated an electric current by sliding a bar magnet into a long wire coil, as shown in Figure 20. Faraday was now changing the magnetic field around the coil by physically moving the magnet. Eleven days later, he attended the Royal Society where he demonstrated that a continuous electric current can be generated by rotating a copper disc attached to two wires between the poles of the 'great horse-shoe magnet' that belonged to the Society. In this case, the magnetic field experienced by the copper disc was varying because the disc itself was moving.

By 4 November, Faraday had demonstrated that simply moving a wire between the poles of a magnet would create an electric current. This is the principle behind the operation of a dynamo. Within three months of inventing the transformer, Faraday had invented a second vital component for the future electrical industry. The function of a dynamo in a power station is to take energy in the form of heat and use it to generate electricity. In the power station, heat derived from sources such as coal, gas or a nuclear reactor is used to boil water to produce steam. This steam then drives a turbine in which a wire coil is rotated in the magnetic field of a huge magnet. The movement of the coil in the magnetic field generates an electric current; mechanical energy in the turbine is converted into electricity.

Faraday had discovered a fundamental symmetry between the electric and magnetic forces, although it had not turned out exactly as he had expected. We know today that an electric current consists of a flow of electrically charged particles, usually electrons. The movement of the electric field produced by these electrons generates a magnetic field.

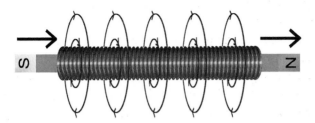

Figure 20 A magnet moving through a coil and generating an electric field, which will cause a current to flow in the coil.

Conversely, as Faraday had now shown, movement of a magnetic field produced by physically moving a magnet generates an electric field that will cause a current to flow, as illustrated in Figure 20.

Michael Faraday is often considered to be the greatest experimentalist in the history of science. He was also a great educator and a promoter of new scientific ideas. In 1825 he established the now famous Royal Institution Christmas Lectures. These lectures for schoolchildren continue to this day and have been presented by some of the world's leading researchers.

The Inverse Square Force Law

Faraday's maps of electric and magnetic fields contain a lot of information. Among other things, they show how the electric force varies with distance. The strength of the electric force is indicated by the density of the field lines in the map. If we look again at a map of the electric field around a charged particle, such as an electron, we can see that the lines of force spread out, which indicates that the force diminishes with distance from the particle. A closer inspection of the map allows us to work out how rapidly the force decreases with distance. Figure 21 shows a two-dimensional cross section of the field.

The diagrams of fields we have looked at so far have been simplified cross sections in two dimensions, but we need to consider the real world of three-dimensional space. First, though, we can imagine that space actually has just two dimensions and see how the electric force would diminish with distance if this were the case. Theorists often invent imaginary universes where the laws of physics are different.

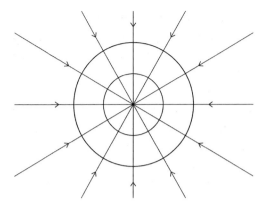

Figure 21 An electric field crossing two concentric circles.

By exploring the physics of a simpler world, we can take a step towards understanding how our own universe works and what is so special about it. The poet W.H. Auden is reputed to have suggested that to understand your own country you need to have lived in at least two others.[20] Physicists feel exactly the same – to understand the physics of our own universe, it is very helpful to investigate the physics of other hypothetical universes. Analysing an imaginary two-dimensional example can help us to understand real physics in three-dimensional space. The two-dimensional example is often simpler to study and can be used as a first step towards solving a real three-dimensional problem. Physics in two dimensions can also be surprisingly different from physics in three dimensions, and this can sometimes offer key insights into the distinctive features of three-dimensional physics.

To help us think about how the electric force varies with distance, we can imagine a two-dimensional world in which an electric field surrounds a charged particle such as an electron, as shown in Figure 21. How does the size of the force change as the distance from the electron increases? If we draw two concentric circles centred on the electron, we can see how the density of the field lines emanating from the electron changes as we move outward from the smaller circle to the larger one. If the radius of the first circle is one metre, then the density of the field lines crossing this circle represents the strength of the electric force that would be felt by a charged test particle one metre from the electron. If the radius of the second circle is two metres, then the circumference of the second circle is twice that of the first circle. All the field lines that cross the first circle also cross the second circle, so the number of lines crossing both circles is the same. But the circumference of the second circle is twice the circumference of the first, so the density of the field lines crossing the second circle is half the density of the field lines crossing the first circle. So at twice the distance from the electron, the field lines have spread out and the strength of the electric force has halved. It is clear from the diagram that at a distance of three metres the force would be reduced to one-third of its value at one metre, and similarly we could work out the size of the force at any other distance. If we measure the strength of the force at a standard distance, such as one metre, we now have a simple relationship that will allow us to calculate the strength of the force at any other distance. Physicists would describe this relationship by saying that the force diminishes with the inverse of distance. But remember, we are considering how the force would behave in an imaginary two-dimensional universe. It would be a mistake to jump to the conclusion that our argument will also apply in three-dimensional space.

Now imagine the field lines around an electron in three-dimensional space. They will radiate outwards, hedgehog-like, in all directions. We can sit our electron at the centre of a sphere with a one metre radius. The strength of the electric force on a charged particle one metre from the electron will now be indicated by the density of the field lines crossing this sphere. If we consider a concentric sphere with a radius of two metres positioned around the first sphere, the surface area of the second sphere will be four times the surface area of the first sphere. All the field lines will cross both spheres, so the number of lines crossing the spheres will be the same. As we double the distance from the electron, the area of the sphere that the field lines are intersecting quadruples and the density of the field lines is therefore reduced to a quarter of its value at one metre. So, if we double the distance between two charged particles, the strength of the electric force is reduced to a quarter of its initial value. We can continue with the same line of reasoning to see that at three times the distance, the force will be one-ninth of its initial value, and in a similar way we can calculate the size of the force at any other distance. This relationship between distance and force is known to physicists as an inverse square force law. It plays a very important role in physics.

Another feature of the electric force that is clear from this line of reasoning is that although the strength of the force diminishes with distance, it never disappears altogether. This is apparent from the diagrams. Our field lines radiate outwards for ever. Their density diminishes with distance, but it never disappears altogether, so although the force becomes weaker with distance it never quite falls to zero. This means that the range of the force is infinite. This property of the electric force is also shared by gravity. Both these forces obey inverse square laws, and so both have an infinite range. They are the only long-range forces that we know of. It is their long-range nature that gives them such an important role in the physics of our everyday lives.

This explanation of how the force operates is based on diagrams first drawn by Faraday. But the diagrams are useful only because they represent the experimental facts well. The inverse square law of the electric field is an experimental fact. It was conclusively demonstrated through measurements that were made by the French physicist Charles-Augustin Coulomb (1736–1806) towards the end of the eighteenth century and published in a set of seven treatises submitted to the French Académie des Sciences between 1785 and 1791.[21]

Maxwell's Beautiful Equations

As we have seen, Faraday's maps give us a succinct way to capture many of the properties of electric and magnetic fields. But to utilise this information effectively it needs to be encapsulated in a mathematical form so that we can use it to perform calculations. Faraday didn't have the necessary mathematical skills to take this step,[22] but another man did: James Clerk Maxwell (1831–79). Maxwell was one of the leading mathematicians of his day. In 1854 he graduated from Cambridge University as second wrangler – the title given to the student who achieves second place in the final exams of the gruelling Mathematics tripos.

Maxwell built on Faraday's experimental work to produce a complete mathematical theory of electromagnetism. He took Faraday's concept of electric and magnetic fields and derived a set of equations that represented these fields mathematically. It is possible to reason from pictures, and Faraday's pictures are a great help in understanding the principles of electro-magnetism, but mathematics is much more powerful. With a precise mathematical description it becomes feasible to extract quantitative predictions that can be tested in the laboratory.

In their usual modern presentation, there are four Max-well equations. The first describes how the electric field permeates space. This equation captures the information in maps such as our map of the electric field around charged particles, shown in Figures 12 and 13. The electric field lines emanate from the positively charged particles and converge on the negatively charged particles. Maxwell's second equa-tion describes how the magnetic field permeates space. This encapsulates maps such as our map of the magnetic field around a bar magnet shown in Figure 14. As far as we know, no magnetic monopoles exist,[23] so all magnetic field lines must form closed loops, and this property is specified in the structure of the second equation.

The third and fourth of Maxwell's equations describe the effects of varying magnetic and electric fields. The third equation is an expression of Faraday's law of induc-tion, which states that a changing magnetic field generates an electric field. This is the effect Faraday had discovered in his experiments in 1831 with the double coil, which is the basis of the construction of a dynamo. The fourth equation encodes the findings of Ørsted and Ampère that a changing electric field or an electric current generates a magnetic field, as so beautifully illustrated in Faraday's experiments with his homopolar motor.

Maxwell's equations capture the results of all Faraday's experimental observations on electromagnetism. The equa-tions are completely general: whatever the distribution of electric charges and magnets, and however these charges and magnets are moving around, the electric and magnetic fields obey Maxwell's equations. From the equations, the evolution over time of the electric and magnetic fields can be calculated, and the trajectories of any charged particles travelling through these fields can be determined. In a sense, by formulating these equations Maxwell achieved for electromagnetism what Newton had achieved for gravitation. He had studied all the experimental results and synthesised a complete theory of electromagnetism that could in principle be used to predict the future behaviour of any electromagnetic system. Maxwell's theory is an elegant unification of the electric and magnetic forces. His beautiful equations provide a mathematical relationship between the electric and magnetic forces and show that these forces are two sides of the same coin. This is the classic example of the unification of forces, and is the model for all the subsequent attempts at further unification of the forces of nature (see Figure 22).

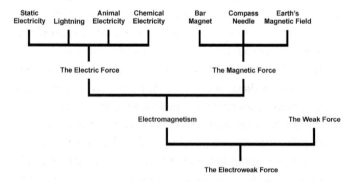

Figure 22 The unification of the electric, magnetic, and weak forces (gravitation and the strong force are omitted).

Let There Be More Light

The truly great theoretical advances in physics seem to give back more than was put into them. Experiments provide clues about the correct form a mathematical description of nature should have, but once this mathematical description has been established it often takes on a beauty and a power all of its own. Probing new equations and seeing what else they are capable of can reveal new secrets about the universe. Maxwell's equations are a superb example of this, and Maxwell himself immediately recognised their power. The equations capture how a changing electric field will induce a magnetic field, and how a changing magnetic field will induce an electric field. This prompted Maxwell to ask whether it was possible to arrange the fields such that the changing electric field would generate a changing magnetic field, and so that this magnetic field would then generate an electric field equivalent to the original electric field. In other words, could the two fields mutually regenerate each other indefinitely, in a self-sustaining manner? Could such mutually self-perpetuating fields exist? Maxwell's equations were the perfect mathematical tools to answer this question.

When Maxwell analysed his equations in 1861 to consider this question, he found that they do indeed lead to just such a solution: an electromagnetic wave, where the electric and magnetic fields oscillate at right angles to each other in a self-perpetuating manner (Figure 23). The waves travel in the direction that is perpendicular to both the electric field and the magnetic field, as indicated by the large arrow in the diagram. But what sort of waves were they? From his equations, Maxwell could determine their speed, and when

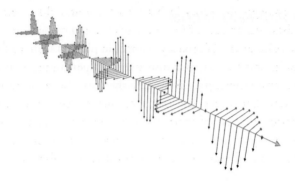

Figure 23 The mutually perpendicular electric and magnetic fields of electromagnetic radiation. The arrows in the horizontal plane represent the magnetic field, the arrows in the vertical plane represent the electric field and the larger arrow shows the direction of motion. The wavelength of the radiation is the distance between two consecutive peaks of the electric field (or of the magnetic field).

he performed the calculation it came out to within a few per cent of the latest measurement of the speed of light. The agreement was so close that it couldn't be a coincidence. This was amazing: Maxwell's theory was telling him that light is an electromagnetic wave. In the process of unifying electricity and magnetism, he had achieved the unexpected bonus of simultaneously uniting both forces with the science of optics.[24]

Maxwell's equations placed no restrictions on the wavelength of electromagnetic radiation, and this seemed to imply that electromagnetic radiation exists over a broad range of wavelengths with visible light making up only a small region of the spectrum. In a series of experiments in the late 1880s, the German physicist Heinrich Hertz provided a conclusive demonstration of the validity of Maxwell's theory. Hertz used an electric current to generate

electromagnetic waves in his laboratory and then detected these electromagnetic waves with a coil of wire on the
other side of the laboratory. Hertz's waves are now known
as radio waves, and within a decade or so of their discovery they were being used for long-distance communication.
We now know that the full spectrum of electromagnetic
radiation ranges, in order of increasing wavelength, from
gamma rays, through X-rays, ultraviolet radiation, visible
light, infrared radiation and microwaves, to radio waves.

Balancing Forces in the World Around Us

How does this all relate to the forces that we feel in our daily
lives? Although gravity is so important to us, it is incredibly weak compared with electromagnetism. We can demonstrate just how weak by taking a small bar magnet and using
it to pick up a metal rod. With the magnet we are able to
overcome the gravitational attraction of the whole Earth on
the metal rod. If we consider the difference in size between
the Earth and the bar magnet, it will give us some idea of the
enormous disparity in the strength of the gravitational force
and the electromagnetic force.

All objects feel the gravitational force. It is always an
attractive force. This means that the gravitational force
can become extremely large if sufficient matter is able to
accumulate into objects such as planets, stars, black holes
and galaxies. By contrast, the electromagnetic force is
sometimes an attractive force and at other times a repulsive force. Particles can have either a positive or a negative
electric charge. Oppositely charged particles attract each
other, but particles with the same charge repel each other.
Whether the electromagnetic force is attractive or repulsive

just depends on the relative charges of the bodies in question. This means that electric charges tend to cancel one another out. If there is a build-up of particles that carry charges of one sign, this will produce an electromagnetic force that will cause oppositely charged particles to flow until the charge is cancelled out or neutralised.

A blinding example of this is a flash of lightning. Friction between tiny ice crystals carried by currents of moist air in the atmosphere can produce a separation of charge between different parts of a cloud and between the base of the cloud and the ground. Eventually the force between the opposite charges becomes so great that there is a flash of lightning. This is simply a manifestation of the flow of charged particles. The separation of charge will then be, at least partially, neutralised by the lightning bolt. So, even though we are surrounded and indeed composed of matter that consists of electrically charged particles, we are not aware of the true strength of the electromagnetic force because most of the opposite charges are matched and therefore cancelled out on microscopic length scales.

Although, fortunately, we do not usually witness the full power of electromagnetism, this force has been known since ancient times and is responsible for everyday phenomena such as static, the magnetic compass, electricity and lightning. What is not so obvious, and became clear only in the twentieth century, is that almost all the forces of which we are aware are due to the electromagnetic force between particles at the atomic scale. A brief visit to the microscopic realm of the atom reveals that atoms are held together by electromagnetic forces. The positively charged nucleus of an atom is surrounded by negatively charged electrons. Collections of atoms form molecules, which are held together by electromagnetic forces. Solids are composed of vast numbers of molecules or huge arrays of atoms that may form a regular crystal lattice; in liquids the bonds between molecules are weaker and there is less organised structure; in gases, the forces between molecules are weaker again. But in all these cases the forces are electromagnetic in origin. The whole subject of chemistry is about the binding of atoms into molecules. All the forces in chemistry are electromagnetic in origin. It is electrons that bind molecules together, and chemical reactions are all about the rearrangement of these electrons.

Let's return to the macroscopic level, to the world we inhabit in everyday life. The reason there is solid ground beneath our feet is that there are strong electromagnetic forces between the atoms in solids. These forces both hold the atoms together, but resist their being pushed too close together. If there were no electromagnetic force holding atoms apart, there would be nothing to resist the pull of gravity and the Earth and everything on it would collapse under gravity. It is not only the objects around us – mountains, houses, cars, chairs, diamonds, raindrops, and so on – that are supported by electromagnetic forces: the same is true of our bodies. The forces we feel when we contract our muscles result from the motion of protein molecules in our muscle cells, so they are also electromagnetic, as are the signals that pass along our nerves informing us of the muscle contractions.

The car provides us with another illuminating example. When we travel in our car, all the forces involved in transporting us around are electromagnetic in origin. The car has a battery which stores electromagnetic energy by maintaining the separation of oppositely charged particles. When the particles are allowed to flow, they produce a current that results in a spark. The spark ignites the petrol, and this explosion pushes on the pistons. The motion of the pistons is transmitted mechanically to other components of the car and ultimately drives the car along. The force produced by the battery is clearly electromagnetic in origin, but so are the other forces. The ignition of the petrol–air mixture is a chemical reaction. In all the interactions involved in the oxidation of the petrol and its conversion into carbon dioxide, electrons are redistributed between different molecules. These interactions are all electromagnetic. The burning of the petrol releases energy which heats the vapour in the combustion chamber. This heat is transmitted between the molecules in the vapour by collisions between the molecules. When the molecules approach one another they are repelled because of the electromagnetic repulsion of their outer electrons. Energy is exchanged between the molecules in these collisions. The increase in the energy of the molecules is interpreted in macroscopic terms as a rise in temperature. Some of the energy of these molecules is now transferred to the molecules in the pistons, again through collisions in which electrons in the vapour molecules and electrons in the metal pistons interact electromagnetically. Similarly, all the mechanical forces that transfer the forces from the pistons through to the axle that drives the car are due to electromagnetic interactions between the electrons in two pieces of metal in contact with each other. And if this were not enough, the electromagnetic force is also responsible for the functioning of the car's headlamps, the car radio and the satellite navigation system.

To summarise, all the forces we are aware of in everyday life have their origins in either gravitation or electromagnetism. The structure we see and feel in the macroscopic world comes about through the interplay between and the balancing of these two forces. The economy in this description of the world is one of the most incredible discoveries in the history of science. However, we will soon be delving into the microcosm, where gravity, because it is so weak, does not play a role.

Further Unification

The construction of a unified theory of the electric and magnetic forces was an amazing achievement of nineteenth-century physics. From the time of Faraday and Maxwell onwards, each of these two forces could no longer be considered in isolation. Maxwell's glorious unified theory of electromagnetism is the model for all modern unified theories. As we will see in the next chapter, the analysis of electromagnetic waves towards the end of the nineteenth century led to quantum theory – the greatest revolution in physics since Newton. This is the great dividing line in physics. All theories of physics that do not incorporate quantum mech-anics are referred to as 'classical'. The physics of Newton, Maxwell and Einstein is classical physics. All modern physics, since the 1920s, is built on quantum theory. This doesn't mean that we now think Maxwell's theory is wrong, but we see it as an approximation to an even better and more accurate quantum theory. Maxwell's theory is the starting point for the quantum theory of electromagnetism. This theory is called 'QED', which stands for quantum electrodynamics. According to quantum theory, light and other electromagnetic waves exist in bite-sized pieces known as photons. We now know that, just as matter is composed of particles such as electrons and protons, so electromagnetic radiation is also composed of particles.

Both the gravitational force and the electromagnetic force are effective over very long distances, which is why we are so familiar with them and why their effects have been studied since antiquity. The middle years of the twentieth century saw the dramatic arrival of nuclear physics. It then became clear that, as well as electromagnetism, there were two hitherto unknown forces operating over very short length scales that are important in the realm of the atomic nucleus. The nucleus is composed of a collection of protons and neutrons. Protons carry a positive electric charge and neutrons are neutral particles, as their name suggests, so they do not carry any electric charge. There is an extremely large electromagnetic repulsion between the positively charged protons within the nucleus, which implies that there must be an even stronger force that holds the nucleus together. This force must be so strong that it can resist the electromagnetic repulsion. It is called the strong nuclear force, or simply the strong force. Unlike electromagnetism, the strong force is a very short-range force. Its effects can be felt only over distances that are smaller than an atomic nucleus. Protons and neutrons both feel the strong force, which is what binds them together in atomic nuclei. Electrons do not feel the strong force, so they are free to roam outside the nucleus.

The other force that is important in the nucleus is a force which is much weaker than electromagnetism. It is called the weak nuclear force or, more usually today, just the weak force. Like the strong force, the weak force is very short-range. The weak force acts on electrons as well as protons and neutrons.

The discovery of the weak and strong forces gave us a total of just four different forces: gravity, electromagnetism, and the weak and strong forces. The interplay between these four forces is responsible for all the complex behaviour we see on

Earth and throughout the universe. It is quite incredible that all the phenomena we see in the universe can be explained by just four forces. Of course, it would be even more remarkable if we could find ways of uniting some of these forces and reducing their number even further. Physicists always seek the simplest possible theories. We can imagine that the most fundamental laws of the universe will be the simplest and most symmetrical possible. Ultimately, it would be wonderful if all the different forces could be reduced to just one.

If we leave aside gravitation, the effects of which are important only over large distances, we are left with three forces that are important in nuclear and particle physics. On the length scale of the atomic nucleus quantum mechanics plays a major role, so any fundamental description of these forces must necessarily be a quantum theory. The problem for modern theorists is to find a quantum theory that unites these forces.

The weak force appears to be very different from the electromagnetic force. For a start, it is much weaker, and it acts over only a very short range. We have seen that the electromagnetic force diminishes with distance according to an inverse square law, so that at twice the distance the force will only have a quarter of the original strength. The electromagnetic force falls away steadily in this way, but it never falls to zero and so has an unlimited range. But the weak force has a range that is very short, much smaller than the size of an atomic nucleus. We will look at the weak force in more detail in a later chapter, but it is clear from these observations that it must operate in a quite distinct way from the electromagnetic force. However, the work of a number of theoretical physicists in the 1960s culminated in an electroweak theory that is designed to unify electromagnetism and the weak force, as shown in Figure 22. This theory is sometimes called the GWS theory, from the initials of Sheldon Glashow, Steven Weinberg and Abdus Salam who were awarded the 1979 Nobel Prize in Physics for their role in the theory's construction.

The main feature of the theory is that at extremely high temperatures the electromagnetic and weak forces are two components of a single force, the electroweak force. The symmetry between the two forces would only be apparent at trillions of degrees, temperatures that could only have occurred at the dawn of the universe in the Big Bang. At lower temperatures the symmetry between the forces is broken; electromagnetism remains a long-range force, but the weak force takes on the characteristics we observe in the laboratory – those of a very weak force that acts over extremely short distances. The Higgs particle is the lynch-pin of the theory, as we will see later. Its discovery by the Large Hadron Collider places the last piece in the jigsaw of the electroweak theory – final confirmation that the unified theory of the electromagnetic and weak forces is correct.

In the next chapter we begin our descent into the heart of matter. This is the realm of the quantum.

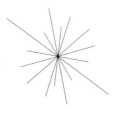

Chapter Three

THE DREAMS THAT STUFF IS MADE OF

On the subject of stars, all investigations which are
not ultimately reducible to simple visual observations
are ... necessarily denied to us. While we can conceive
of the possibility of determining their shapes, their sizes,
and their motions, we shall never be able by any means to
study their chemical composition or their mineralogical
structure ... Our knowledge concerning their gaseous
envelopes is necessarily limited to their existence,
size ... and refractive power, [and] we shall not at all be
able to determine their chemical composition or even their
density ... I regard any notion concerning the true mean
temperature of the various stars as for ever denied to us.

Auguste Comte, *Cours de Philosophie Positive* (1835)[1]

Sociology and the Stars

The nineteenth-century French philosopher Auguste Comte is generally considered to be the founder of sociology as a scientific discipline. He believed firmly in the principles of the Enlightenment and the egalitarianism of the French Republic, and felt that the scientific method should be utilised for the benefit and improvement of society. He is also regarded as the most influential proponent of the philosophy of positivism. In Comte's view, the only genuine knowledge available to humanity was that which was gained by direct observation and the scientific method. All other sources of knowledge, whether derived from divine inspiration, religious faith or traditional authority, he considered worthless. Science was the only route to the truth; all else was meaningless metaphysical speculation. Comte presented his ideas to the world in his 'Course in Positive Philosophy', published in several volumes between 1830 and 1845. In this work he discussed what knowledge would, in his opinion, always be beyond the reach of humankind. He gave an example of such knowledge in the quotation at the start of this chapter. In Comte's considered opinion, the temperature and chemical composition of the stars would always remain beyond our reach.

Comte's example may appear quite reasonable, as the distance to the nearest star other than the Sun is over four light years. This is a vast distance. The stars are like tiny atolls in great oceans of space. Light travels through space at almost precisely 300,000 kilometres per second. At this speed it takes light eight and a half minutes to reach us across the 150 million kilometres that separate us from the Sun. In four years, light will travel an immense 40 trillion (million

million) kilometres. On the human scale this is an almost inconceivable distance, and it shows just how far it is to even the nearest star.

Comte must have thought that he was stating the obvious when he made his pessimistic pronouncement. It was clear to everybody that it would never be possible to travel to Betelgeuse, Rigel or even the Sun with a thermometer and a chemical laboratory. However, any good modern astronomy guidebook will tell us the temperatures of hundreds of naked-eye stars. We will also be able to read about which chemical elements the star is composed of and how abundant these elements are. For instance, the temperature of the outer layers of the Sun is about 5,800 K. (Although the Centigrade scale we commonly use is convenient for discussing everyday temperatures, the Kelvin scale of temperature is more natural. Zero degrees Kelvin corresponds to the absolute zero of temperature, and 0 °C is equal to 273.15 K, so all the temperatures on the Kelvin scale are just over 273 degrees higher than their Centigrade equivalents.) According to the section of our guidebook on the stars of the constellation Orion, the surface temperature of bloated Betelgeuse is about 3,500 K, and the temperature of the brilliant blue Rigel is about 11,000 K. We can also read in our guidebook that the main chemical constituents of all stars are the two lightest elements, hydrogen and helium. Stars contain smaller quantities of many other elements as well, such as carbon, oxygen, neon, silicon and iron. But how do we know this?

Despite Comte's conviction that it was impossible, determining the temperature of the stars is relatively easy. We know from everyday experience that when we dramatically raise the temperature of a metal object its colour will change. In the medieval world, the technology of the blacksmith was shrouded in an aura of magic. An important part of the blacksmith's art was the ability to judge the temperature of the glowing hot metal that he was working so that each step in the forging process could be performed at the correct temperature. The best way to judge the temperature of the metal was by its colour. A rod of steel heated to about 1,000 K will glow red. As its temperature is increased further, it will glow orange, and at even higher temperatures it will turn yellow, white and then blue.[2] The rod is actually emitting radiation even before it is heated to such high temperatures. At temperatures below 1,000 K we may not see the rod glow at all, but it is emitting electromagnetic radiation whose wavelength is longer than red light. This infrared radiation is invisible to us, as it is outside the range our eyes are sensitive to, but we can feel its effect as heat if we move our hand towards the rod. The Sun, whose surface temperature is about 5,800 K, emits some of its radiation as visible light, but most of its radiation has a wavelength that is shorter than that of violet light. This is known as ultraviolet radiation. It is also not visible to us, but it is high-energy radiation and harmful to human eyes, which is why it is dangerous to look directly at the Sun. It is clear, then, that there is a close relationship between the temperature of an object and the light the object emits. All objects emit electromagnetic radiation over a broad range of wavelengths at all temperatures. But as the temperature of an object rises, the typical wavelength of the radiation it emits decreases, and we are aware of this because we see the colour of the object change.

This simple correlation enables us to determine the temperature of an inaccessible object just by analysing its colour, even if that object is as distant as the stars. Since the end of the nineteenth century, the colour of stars has been the most important property used in their classification. The colour of a particular star's light enables astronomers to work out how hot it is, and its temperature can then be used to determine many other significant features of the star. From our theoretical understanding of stellar evolution, we can relate the temperature of a star to its mass and the stage it has reached in its life cycle. We know that blue giants like Rigel are very hot young stars, yellow dwarf stars have a similar temperature to the Sun's, and red supergiants such as Betelgeuse have swollen to enormous dimensions in the later stages of their life. The colour of some of the brightest stars, like Betelgeuse, is obvious even to the naked eye. Betelgeuse, which forms Orion's left shoulder as we look at it, has a distinct orange colour. If this vast, diffuse star took the place of the Sun in our solar system, it would easily contain the Earth's orbit, and its outer regions would reach as far as Jupiter.

One of the traditional delights of amateur astronomers is to observe the delicate shades of colour of multiple stars. These star systems consist of two or more stars that orbit around one another. Multiple stars with contrasting colours can be striking when viewed through a telescope. One beautiful example is the star Albireo that forms the head of Cygnus the Swan, the crucifix-shaped constellation that is prominent in summer skies in the northern hemisphere. Even a small telescope will show that Albireo is a stunning double star system that consists of a bright yellow star and a fainter blue companion.

The Cosmic Barcode

Clearly, Comte was quite mistaken in his assertion that the temperature of the stars could never be known. Despite his great belief in science, he had seriously underestimated the ingenuity of scientists. His doubts about the possibility of knowing the chemical make-up of the stars were also unfounded. Just two years after his death, in 1857, a whole new world of chemistry was opened up by a sensational discovery made by two German scientists, Robert Bunsen and Gustav Kirchhoff. An ecstatic Bunsen excitedly announced to a colleague:

> At present Kirchhoff and I are engaged in a common work which doesn't let us sleep ... Kirchhoff has made a wonderful, entirely unexpected discovery in finding the cause of the dark lines in the solar spectrum ... Thus a means has been found to determine the composition of the sun and fixed stars with the same accuracy as we determine sulphuric acid, chlorine, etc., with our chemical reagents. Substances on the earth can be determined by this method just as easily as on the sun, so that, for example, I have been able to detect lithium in twenty grams of sea water.[3]

Bunsen and Kirchhoff had discovered that each chemical element has a unique fingerprint that would allow them to track down its presence anywhere in the universe. If an element is heated in a flame until it glows, the light it emits has a characteristic colour. And when that light is passed through a prism, the resulting spectrum appears as a number of bright lines corresponding to light at a series of sharply defined wavelengths. For instance, sodium will burn with an intense yellow glow. Many street lamps contain sodium vapour, which is why they shine with a yellow hue. The light that sodium emits, when passed through a prism, splits into two sharp lines in the yellow region of the spectrum. Bunsen and Kirchhoff took each element in turn and heated it in a flame until it glowed. They then passed this light through a prism and recorded its unique fingerprint. The gas-fuelled burner that had been invented by Michael Faraday was invaluable for their work. Peter Desdega, one of Bunsen's assistants, further improved the design of the burner and then marketed it very successfully under Bunsen's name.[4] It is now the best-known piece of equipment in the school science laboratory.

The two elements that are chemically most similar to sodium are lithium and potassium, which are the elements immediately above and below sodium in the periodic table. The flame test is a very simple way to distinguish these two similar metals from sodium. When heated in the flame of a Bunsen burner, potassium glows with a characteristic lilac-coloured flame that is occasionally seen in a log fire, while lithium burns with a bright red flame. The range of colours that can be produced in this way is very useful in the manufacture of fireworks, whose brilliant colours are achieved by mixing a variety of metallic elements with the highly flammable powder that is ignited when the firework burns. If you are curious about the origin of the colours the next time you see a Roman candle or sky rocket, then here is the recipe: lithium or strontium compounds are added to produce a bright red glow, calcium is used for an orange glow, sodium for yellow, barium for green and copper compounds for blue.

Bunsen and Kirchhoff's critical insight was to match the dark lines in the spectra of the Sun and the stars to the lines they were observing in their laboratory. That way they could tell which elements were present in which star. Each element has a unique spectral fingerprint. A glowing sample of the element emits light at a characteristic set of wavelengths that can be recorded in the laboratory as what is called an emission spectrum. If the same element is present in the atmosphere of a star, it will absorb the light streaming out from its core at exactly the same wavelengths, resulting in what is called an absorption spectrum. The star's light is thus depleted at these wavelengths, which is why they appear as dark lines in the star's spectrum. So the dark lines in the star's spectrum caused by the presence of a particular element are at exactly the same wavelengths as the bright lines that form that element's spectrum in the laboratory. A star's spectrum is like a cosmic barcode containing precise information about the identity of each element that is present in the star's atmosphere. These telltale signs of each element are clear even at a distance of trillions of kilometres.

Plate 4 shows an example of this correlation. At the top are certain dark lines that appear in the spectrum of the Sun. Below this are schematic spectra of four elements: sodium, mercury, lithium and hydrogen, as would be produced in the laboratory. The dark lines in the solar spectrum do not match up with the lines in the spectra of mercury and lithium, but they do match up with lines in the spectra of sodium and hydrogen. So we can deduce that hydrogen and sodium are present in the Sun, but mercury and lithium are not present in significant quantities.

By using this technique it was even possible to discover new elements in the stars. Helium is an inert gas – it doesn't form chemical compounds. It also has a very low mass. For these reasons any helium that gets released from within the Earth, in an event such as a volcanic eruption, will rapidly escape from the Earth's atmosphere into space. We know now that helium is the second most abundant element in the universe.[5] Yet despite this cosmic abundance, helium remained unknown to science until the middle years of the nineteenth century. During a solar eclipse in 1868, the French astron-omer Pierre Janssen recorded the spectrum of the ghostly glow around the Sun, known as the corona, which is visible only during a total eclipse. He found a dark absorption line that could not be identified with any line in the emission spectrum of any known element. Later that same year, the British astronomer Norman Lockyer confirmed Janssen's discovery from a spectrum he obtained when observing the Sun through the haze of London's smoky sky. Lockyer correctly deduced that the absorption line was caused by a hitherto unknown element. He named the element 'helium' after Helios, the Greek god of the Sun. Lockyer mistakenly assumed that his new element must be a metal, which is why he gave it a name that contains the suffix '-ium' that is usually reserved for metallic elements.

Spectroscopy, as the discipline is called, became an invalu-able tool for the chemists of the nineteenth century, but they could not explain the physical origin of spectral lines. Why each element emitted and absorbed light at its own char-acteristic wavelengths remained a mystery. The solution lies within the atom. Later investigators who probed the struc-ture of atoms followed a winding trail of cryptic clues that would finally reveal the answer.

The Atom

It seems surprising today, but Dmitri Mendeleyev was never convinced that atoms really exist. Despite the advantages of the atomic hypothesis, many scientists remained scep-tical about the existence of atoms as physical entities until the early years of the twentieth century, when the evidence for them finally became overwhelming. This reluctance to accept the reality of atoms may have come partly from their association with the speculations of ancient Greek philosophers. But the atoms of the modern chemist and physicist are quite different from the indivisible particles that Democritus had envisaged. We know today that atoms are in turn composed of even smaller particles, and the discovery of these even smaller components, such as the electron, was important in overcoming the final resistance to the acceptance of the physical existence of atoms. Although atoms represent a very important level in the organisation of matter, they are certainly not the ultimate indivisible constituents of matter.

In 1884, Joseph John Thomson, known to his colleagues as 'J.J.', was appointed director of the Cavendish Laboratory in Cambridge, and over the next decade he built the laboratory into one of the world's leading centres for physics research. In a series of experiments he conducted there in the 1890s, he analysed the properties of electrically charged rays that were produced by a new piece of equipment known as a cathode ray tube. This device would become the main component of the television set: the image on older TV screens is produced when these cathode rays strike the back of the screen. But Thomson was not concerned with possible future applications of his apparatus. He was seeking to understand the structure of matter and, in particular, the identity of the rays.

By 1897 Thomson had measured the mass and the electric charge of the negatively charged particles of which the rays were composed, and surmised that this particle was a fundamental component of atoms. It was the first subatomic particle to be identified. Thomson referred to it as a 'corpuscle', but we know it today as the electron. With the discovery of the electron, few scientists could deny the existence of atoms any longer. But as yet there was no experimental evidence for how electrons might be organised within the atom, or even how many electrons an atom might contain. Thomson himself suggested that the atom consisted of negatively charged electrons held together in a positively charged amorphous matrix. This picture was given the deliciously Victorian nickname of the 'plum pudding' model of the atom. The electrons were the plums held together in the duff of the pudding.

In 1895 Thomson was joined in Cambridge by a young New Zealander who was to become the greatest experimental physicist of the twentieth century. The big, burly Ernest Rutherford (1871–1937) was an unlikely figure to achieve such greatness. He had been raised on a remote farm in late-Victorian New Zealand, but had won a scholarship that offered the opportunity to study with J.J. Thomson in the leading scientific laboratory in the British Empire. Meanwhile, in Paris in 1896, Henri Becquerel discovered radioactivity. This completely unexpected breakthrough would launch the era of nuclear physics. After a brief period studying the possibilities of radio transmission, Rutherford took up the challenge of understanding radioactivity, and he soon established that there were two different types of ray being emitted by the radioactive elements he was studying. As he didn't know the identity of these rays, he named them alpha rays and beta rays. Rutherford would later prove that the rays are streams of particles, and an alpha particle is identical to a helium atom stripped of its electrons.

Rutherford realised that these high-energy alpha particles shot out by radioactive materials were the perfect missiles with which to probe the structure of matter. Using a highly radioactive radium source, Rutherford, who had by now been recruited for the new and generously funded McGill University in Canada, investigated the effect of bombarding thin sheets of gold foil with alpha particles.[6] The reason for using gold foil was that gold is a very malleable metal, and it is possible to hammer it into extremely thin sheets. This was important because it ensured that any deflection in the path of an alpha particle would almost certainly be due to a single encounter with a gold atom. Also, gold foil was readily available as it was widely used in jewellery and decoration. Rutherford hoped that the impact of the high-energy alpha particles with the atoms in the gold foil might offer some clue to the structure of the atoms. After passing through the foil, each alpha particle would hit a zinc sulphide screen and the impact would produce a tiny flash of light. Even the impact of a single alpha particle could be detected in this way. Rutherford found that the narrow beam of alpha particles that he fired at the foil was spread slightly by its passage through the foil. In other words, some of the alpha particles underwent small deflections as they passed through. This suggested that there was some interaction between the alpha particles and the atoms in the foil, but that this interaction was quite small, as Rutherford expected.

Even at this early stage in his career, Rutherford was being described by his colleagues as a 'force of nature'. A classics professor called John McNaughton who generally dismissed scientists as 'plumbers' and 'destroyers of art' described his first encounter with Rutherford:

> We paid our visit to the Physical Society. Fortune favoured us beyond our deserts. We found that we had stumbled in upon one of Dr Rutherford's brilliant demonstrations of radium. It was indeed an eye-opener. The lecturer himself seemed like a

large piece of the expensive and marvellous substance he was describing. Radioactive is the one sufficient term to characterise the total impression made upon us by his personality. Emanations of light and energy, swift and penetrating, cathode rays strong enough to pierce a brick wall, or the head of a Professor of Literature, appeared to sparkle and coruscate from him all over the sheaves. Here was the rarest and most refreshing spectacle – the pure ardour of the chase, a man quite possessed by a noble work and altogether happy in it.[7]

In 1907 Rutherford accepted a position at the University of Manchester. He felt that a return to Britain would allow him to interact more easily with the leading scientists of the day. In Manchester he continued the cutting-edge research he had begun in Canada. He set a young researcher called Hans Geiger the task of repeating the gold foil experiment to obtain accurate measurements of the extent of the deflection suffered by the alpha particles when they passed through the gold foil. It was a painstaking task: Geiger had to sit in the dark, staring through a microscope at a tiny square marked on the zinc sulphide screen and count the number of faint flashes from the alpha particles that were hitting the square. After repeating the procedure for a range of angles of deflection and counting thousands of scintillations, he was able to report to Rutherford that the number of scattered alpha particles dropped significantly as the angle of deflection increased, and that beyond a few degrees of deflection there were no alpha particles.

Rutherford now made a remarkable suggestion. He proposed that they should check whether any of the alpha particles were being reflected back by the gold foil. On the face of it, this was a very peculiar idea because all the alpha particles that had been observed so far had ploughed on through the foil with barely any deflection at all, which seemed to agree with common-sense expectations. Rutherford and Geiger agreed that this apparently unpromising project should be given to a young student called Ernest Marsden, who had recently arrived from New Zealand. Just a couple of days later, a very excited Geiger came to see Rutherford again. He reported that he and Marsden had performed the experiment and had actually found that a very small proportion of the alpha particles were indeed being reflected backwards. Rutherford was gobsmacked. He later recalled that 'It was quite the most incredible event that has ever happened to me in my life. It was almost as incredible as if you fired a fifteen inch shell at a piece of tissue paper and it came back and hit you.'[8]

Rutherford saw clearly what the implications of Geiger and Marsden's experiment were. Here was an unmistakable message about the structure of the atom. Each alpha particle that rebounded from the gold foil must have collided with a gold atom and been completely reflected backwards. These large deflections could not be the result of alpha particles colliding with the electrons in the gold atoms, because electrons are far too light and would simply be brushed aside by the high-energy alpha particles. It was also clear that the mass of the atom could not be spread out in an amorphous matrix, as in the plum pudding model, because that would not be sufficiently resilient to bounce an alpha particle straight back. The only way that a large deviation in the flight of an alpha particle could be produced was if virtually all the mass of the atom were concentrated into a very tiny region, an extremely dense nucleus. The large deflections occurred on the rare occasions when an alpha particle underwent an almost head-on collision with this nucleus and ricocheted away in a completely different direction.

In order to test this idea thoroughly, Rutherford derived a mathematical formula to describe the scattering of an alpha particle, assuming that the force between the alpha particle and the atomic nucleus was electromagnetic in origin. Geiger and Marsden then conducted a series of experiments to test the validity of Rutherford's formula. Each factor in the formula was checked by reconfiguring the experimental set-up numerous times with different parameters. The experiment was repeated with different metal foils, and the thickness of foil was also varied.[9] In all, Geiger and Marsden counted hundreds of thousands of scintillations produced by alpha particles hitting their zinc sulphide detector. Rutherford's formula fitted the experimental findings perfectly. From the formula, they could determine the size of the atomic nucleus. The diameter of a gold atom is about 0.3 nanometres (3×10^{-10} m) and there are a million nanometres in a millimetre, but the nucleus of the gold atom is a mere one-twenty-thousandth of the diameter of the atom (1.5×10^{-14} m).[10] If we scale the atom to dimensions we are familiar with, the size of the nucleus in relation to the whole gold atom is comparable to that of a marble in the volume of the world's largest church, St Peter's Basilica in Rome.

Rutherford announced the discovery of the nucleus to the Manchester Literary and Philosophical Society on 7 March 1911, just over a hundred years after John Dalton had first presented his theories about atoms to the same esteemed society.[11] Rutherford now proposed a new model of the atom: an extremely small and dense positively charged nucleus surrounded by a collection of electrons whizzing around it, like a miniature solar system.

Quantum Mechanics

Rutherford's discovery of the nucleus was a tremendous step forward in understanding the structure of the atom. But his 'planetary model' of the atom posed a serious problem for established physics. In the solar system, the gravitational attraction between the Sun and the planets holds the planets in orbit around the Sun. Similarly, in Rutherford's atom the electromagnetic attraction between the negatively charged electrons and the positively charged nucleus would hold the electrons in their orbits around the nucleus. However, Maxwell's equations show that the motion of an electron around the nucleus would cause it to continuously emit electromagnetic radiation.[12] Its energy would very quickly dissipate, causing it to spiral rapidly into the nucleus. Rutherford's planetary atom would be stable for no more than a fraction of a second. The solution of this puzzle would require the development of quantum mechanics, the biggest revolution in physics in the 250 years since the publication of Newton's *Principia*.

The German physicist Max Planck had fired the first shots of the quantum revolution a decade earlier. Planck was a talented musician from a traditional middle-class academic background, but he chose a career in physics. By the final decade of the nineteenth century he had become a very experienced and well established theoretical physicist, and held a professorship at the University of Berlin. Planck agonised over how matter emits light, and specifically over how much light an object emits as its temperature changes. There was a serious problem with the calculations, which were giving a crazy answer, but eventually the stubborn experimental facts forced Planck, who was conservative by nature, to spark the revolution. The quantum physicist Max Born described Planck in these words:

> He was by nature and by the tradition of his family conservative, averse to revolutionary novelties and sceptical towards speculations. But his belief in the imperative power of logical thinking based on facts was so strong that he did not hesitate to express a claim contradictory to all tradition, because he had convinced himself that no other resort was possible.[13]

Planck found that he could cure the problems with the calculation and derive the correct relationship, but only by making a radical assumption that clashed with the classical physics of Newton and Maxwell. Planck called his explanation the 'quantum' hypothesis. At the time he described it as 'an act of despair', but he later recalled that he 'was ready to sacrifice any of my previous convictions about physics' to solve the problem.[14]

The term quantum literally means 'portion' and is derived from the Latin for 'how much'. (The plural is quanta.) Planck originally applied this term to the oscillations within a solid body, such as a crystal, but the quantum revolution really gathered pace in 1905, when Einstein explained an experimental result – the photoelectric effect – by reasoning that light itself is lumpy rather than continuous. This means that, although in many respects it behaves like a wave, light is actually composed of particles. We call these particles of light photons. However, we must not forget that light also has wave-like properties – as we saw in the last chapter, it can be explained as an electromagnetic wave. Sometimes photons are described as 'wave packets'. The amount of energy each photon carries depends on the wavelength of the light: the shorter the wavelength, the higher energy. Photons of red light therefore carry less energy than photons of blue light. It was for this work that Einstein was awarded his Nobel Prize in Physics in 1921, and not for his more famous work on relativity.

The relationship between the wavelength of a wave and the energy that it carries makes sense intuitively. Imagine tying one end of a skipping rope to a tree. Shake the rope to generate waves. It doesn't take much effort to send a wave with a long wavelength off towards the tree. But in order to send waves with a short wavelength on their way, you have to shake the rope vigorously and put more energy into the waves. This is true in general: a wave with a short wavelength and a high frequency carries more energy than a wave with a long wavelength and a low frequency. Planck introduced a new fundamental constant into physics that determines the length scale at which quantum effects become important. In effect, this constant – which is known as Planck's constant – is a measure of the 'granularity' of the universe. It allows us to convert between the wavelength of a photon and the amount of energy it carries.[15] Because Planck's constant is tiny, quantum effects usually only show up when considering extremely small objects such as photons or atoms. In a hypothetical universe where Planck's constant was much larger, objects that we are familiar with on the human scale, such as cars, trees, houses and people, would all be associated with fluctuating waves. The classical certainty that we take for granted in our interactions with the world would be lost in a hazy miasma.

In 1912, a young Danish theoretical physicist, Niels Bohr, came to Manchester to work alongside Rutherford. Bohr recognised Rutherford's immense stature as an experimental physicist and knew that the results of his experiments had to be taken seriously. If established physics could not accommodate atoms with the structure that Rutherford's experiments implied, then it would take daring new theoretical ideas to explain them. Bohr borrowed the strange new quantum hypothesis of Planck and, with a flash of genius, adapted it to explain how electrons behave in Rutherford's planetary atom. Bohr could give no rationale for his novel methods, but by making this radical departure from classical physics, he produced the first model of the atom to offer precise predictions that could be tested in the laboratory. It would not be fully appreciated for another decade, but the reason for the success of Bohr's model was that it treated electrons as waves. The electrons were not orbiting the atomic nucleus like planets orbiting the Sun. In Bohr's model they formed waves around the nucleus, and they could exist in the atom only if an exact number of wavelengths fitted into their orbit around the nucleus. Quantum mechanics was first conceived when Planck and Einstein showed that although light behaves like a wave, it is actually composed of particles. Now electrons, which appeared to be particles, were shown to have wave-like qualities. This bizarre wave–particle duality, as it is called, lies at the heart of quantum mechanics and is the reason why quantum mechanics represents such a radical departure from classical physics. In order to see how electron waves help us to understand the structure of the atom, we must take another trip back to ancient Greece.

A Musical Interlude

The ancient Greek philosopher Pythagoras of Samos, famous for his theorem about right-angled triangles, is also credited with having discovered the relationship between the length of a string and the pitch of the musical note that it produces when it vibrates. There are various ways to set up vibrations in the string of a musical instrument: it may be plucked, bowed or hit with a hammer. To alter the pitch of the note, the length of the string that is vibrating must be altered: the shorter the string, the higher the note. A guitarist or violinist will place a finger at the appropriate position on the fretboard to shorten a string and raise the pitch of the note by the desired amount.

Musical scales are collections of notes with simple relationships between the lengths of their waves. Only then do the notes sound harmonious when played together. For instance, if the length of the vibrating string is halved, thereby halving the wavelength of the vibration, the frequency of the vibration is doubled, and the note we hear is an octave higher. Halving the length of the string again raises the note by a further octave.

The 'note' we are referring to here is the fundamental mode of vibration of the string, known as its first harmonic. This is the principal tone we hear when a string of an instrument is played, but the string's vibration will always include higher harmonics as well. In effect, even when a single string is played we simultaneously hear a collection of harmonious notes. This is what gives a musical instrument its tone, or timbre. The proportion of each of the higher harmonics depends on the way the vibration in the string is produced; this is why a violin sounds different depending on whether it is plucked or bowed, and why a violin sounds different to a piano.

The fundamental mode of vibration is shown at the bottom of Figure 24. Exactly half of one wavelength of this mode fits on the string. The wave shown above it is the second harmonic of the string. One complete wavelength of the second harmonic fits on the string. In musical language, if the first harmonic produces the note C, the second harmonic will also produce the note C, but one octave higher. Shown above this wave is the third harmonic. In musical terminology this mode produces the note G, which is a fifth above the second harmonic. (G is the fifth note in the scale of C.) At the top is the fourth harmonic of the string. Its wavelength is one quarter of the wavelength of the first harmonic and produces a note that is two octaves above it. When a string is plucked it will simultaneously vibrate in all the different ways shown in Figure 24, and will produce even higher harmonics too. Most of the energy in the vibration will be in the fundamental mode, and this is the main note that we will hear, but we hear the other harmonics as well.

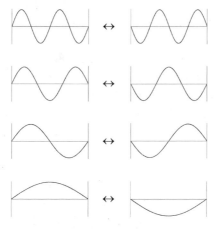

Figure 24 Modes of vibration of a string. Each of the four pairs of diagrams shows two momentary snapshots of a vibration, at opposite points in the cycle. The full vibration is an oscillation between these two positions. The bottom pair of diagrams shows the fundamental mode of vibration, also called the first harmonic. Above it are the second, third and fourth harmonics.

The arithmetical relationship between the wavelength of a vibration and the musical note we hear is a simple rule, but it captures the essence of how changing one quantity (the length of a vibrating string) is related to changes in another quantity (the pitch of the note it produces). It was one of the first scientific laws to be discovered, and as such it became the model for other laws. The fact that the physical relationship between these two quantities could be captured in a simple numerical pattern had a profound significance for Pythagoras, who saw it as a deep insight into the orderliness of the universe. The regularities in the motions of the heavenly bodies suggested that they too could be explained by similar elegant laws. This analogy was the basis for the ancient belief in 'the music of the spheres', an idea that remained influential throughout the Middle Ages, when the four liberal arts – arithmetic, geometry, music and astronomy – comprised the quadrivium, the syllabus of the mediaeval university. Even early scientists, such as Kepler, sought musical relationships in the motions of the planets.

The Music of the Spheres

As it turns out, 'the music of the spheres' is a much better metaphor for the description of electrons in an atom than for the orbits of the planets, because electrons really do behave like waves. The best way to see the implications of this is to consider an extremely simple model, of an electron in one dimension – a line. In this model the entire universe of the electron consists of a segment of a line, and the electron can move only in this one dimension. Another way of looking at the electron's world is to imagine that there is a barrier at each end of the line segment that prevents the electron from escaping, no matter how much energy it has. Despite the extreme simplicity of this one-dimensional model, it can illustrate many of the important features of quantum mechanics.

According to quantum mechanics, the electron is associated with a wave. In our one-dimensional model this wave must fit exactly between the two barriers at the ends of the line segment. Only certain waves will fit between these barriers, so the electron's universe is similar to a guitar string. The possible waves that may be formed by the electron wave are the same as the harmonics of the guitar string, as illustrated above. And as with the guitar string, each of these waves represents one of the ways in which the electron wave can vibrate. There is a simple relationship between our electron's energy and its mode of vibration: the shorter its wavelength, the higher its energy. The energy goes up in steps, like the rungs of a ladder. The first harmonic has the lowest energy and forms the lowest rung of the ladder. In this wave mode the electron has 1 unit of energy. An electron forming the second harmonic wave, one octave higher, is on the second rung of the energy ladder and has 4 units of energy. On the third rung of the energy ladder it has 9 units of energy. And so on. The energy goes up, increasing with the square of the energy level, as the electron rises up the ladder. This model has already revealed one of the famous features of quantum mechanics, namely that our electron can exist only in a discrete set of energy levels. In other words, there are only certain values of energy that the electron wave can have, and it is impossible for it to have an energy that falls in the gap between any pair of energy levels. In classical physics we would think of the electron as a particle with no restriction on its energy, but in our simple quantum model, the electron forms a wave and its energy can only take certain values: 1 unit, 4 units, 9 units, 16 units, or any square number of units.

All physical systems have a tendency to sink towards the lowest energy state that is available. This is just like water flowing downhill to reach the sea. So an electron in an atom will tend to fall into a lower energy level. The electron can lose energy by emitting radiation. So if the electron is on, say, the fourth rung of the energy ladder, with 16 units of energy, it can emit a quantum of light, a photon, and fall down to one of the lower rungs. The energy of the emitted photon must equal the difference in energy between the two rungs. If the electron starts on the fourth rung and falls to the third rung, the photon it emits will carry away $16 - 9 = 7$ units of energy. If the electron starts on the fourth rung and falls to the second rung, the photon will carry away $16 - 4 = 12$ units of energy. If the electron falls all the way from the fourth rung to the lowest rung, the photon will be emitted with $16 - 1 = 15$ units of energy. Our electron is now displaying another famous feature of quantum mechanics: a discontinuous change in energy known as a quantum jump or quantum leap, a term that has slipped into everyday usage.

We can now see a further very important quality that is quite general for all quantum systems. The electron can continue losing energy by emitting photons only until it has fallen down to the lowest rung of the energy ladder, where it has just one unit of energy. It now has the lowest energy it can possibly have, and is in what is called its ground state. The electron cannot lose any more energy, as there is no lower energy state for the electron to fall into. This was the first triumph of quantum mechanics: to explain why stable atoms exist and electrons do not spiral away down into the nucleus.

Quantum Barcodes

Bohr realised that his analysis of the atom could also explain the origin of the lines in the spectra of the elements and thereby solve the puzzle of the cosmic barcodes. In his model, the electron waves wrap themselves around the nucleus to form circles and ellipses, rather than the straight-line segments of our simple one-dimensional model, so the exact details are different and the rungs on the energy ladder are spaced out in a different way, but the general features are just the same. Bohr explained the lines in the spectrum of an element as follows. When the element is heated, its electrons are excited into higher energy levels. When it cools, those electrons fall back into lower energy levels and emit photons whose energy is precisely equal to the difference between the two energy levels. This is the origin of the bright emission spectrum observed in the laboratory, which is produced by the light emitted by electrons as they drop from one energy level to a lower one. When the element is present in the atmosphere of a star, its atoms are bathed in the light radiated by the star, and their electrons will absorb light of just the right energy to boost them into higher energy levels of the atom. This depletes the light from the star at these wavelengths and produces the dark lines in the star's absorption spectrum.

Bohr's model was particularly easy to apply to the simplest atom, hydrogen, as it contains just a single electron. Using his model, he could quite easily calculate the size of the gaps between all the energy levels in the hydrogen atom. So, not only could he account for the general features of the hydrogen line spectrum, he also computed the exact wavelength of the light that produced each line. When compared with laboratory measurements, Bohr's calculations matched every known line in the hydrogen spectrum. That was remarkable enough, but in addition his model predicted whole families of other lines that had never been observed because they were outside the visible spectrum. These lines were all duly found in the laboratory, in the ultraviolet and infrared regions of the spectrum of hydrogen. For this outstanding work, Niels Bohr was awarded the 1922 Nobel Prize in Physics.

It is much more difficult to calculate the energy levels of the electrons in atoms that are more complicated than hydrogen, even atoms as simple as helium, the second element in the periodic table. The helium atom contains two electrons, and each electron will occupy one of the energy levels in the atom. But with more than one electron in the atom, the mutual repulsion between the electrons complicates the analysis and alters the gaps between the energy levels. However, even without undertaking such calculations, it is clear that the spectrum of the helium atom will be different to the spectrum of the hydrogen atom, and that in general the electrons in each type of atom will have their own set of discrete energy levels. The atoms of each element in our one-dimensional world are like the diverse stringed instruments of the orchestra. Element number one might be a double bass, element number two a cello and element number three a viola. Just as the range of notes that can be played on each instrument differs, so each atom has its own unique set of energy levels available for its electrons. This means that each element will have its own characteristic set of wavelengths at which it will absorb and emit photons, just as Bunsen and Kirchhoff had discovered fifty years earlier. The fundamental origin of the cosmic barcodes was no longer a mystery.

The Scanning Tunnelling Microscope

'Why did your spear sting when the point was a half foot
away from where it made me bleed?'

'That spear,' he answered quietly, 'is one of the first
things that I ever manufactured in my spare time ... '

'The point is seven inches long and it is so sharp and thin
that you cannot see it with the old eye. The first half of the
sharpness is thick and strong but you cannot see it either
because the real sharpness runs into it and if you saw the one
you could see the other or maybe you would notice the joint.'

'I suppose it is far thinner than a match?' I asked.

'There is a difference,' he said. 'Now the proper sharp
part is so thin that nobody could see it no matter what light
is on it or what eye is looking. About an inch from the end
it is so sharp that sometimes – late at night or on a soft bad
day especially – you cannot think of it or try to make it the
subject of a little idea because you will hurt your box with
the excruciation of it.'

Flann O'Brien, *The Third Policeman* (1940)[16]

We should pause for a moment to consider just how small
an atom is. The average distance between the electron and
the nucleus in a hydrogen atom is about one-twentieth of
a nanometre. This gives us a rough figure for the diameter
of a hydrogen atom of about one-tenth of a nanometre. The
diameter of a carbon atom is about twice this size. Even the
biggest atoms, such as gold atoms, are not much larger. The
diameter of a gold atom is about three times the diameter
of a hydrogen atom. But what does this mean in everyday
terms? There are ten million nanometres in a centimetre:
one hundred million carbon atoms in a line would stretch

a distance of just two centimetres. In a cube with edges two centimetres long we could pack one hundred million, times one hundred million, times one hundred million carbon atoms. So a cubic block of carbon of this size, which we could easily hold between our fingers, whether it is a priceless diamond or just a humble piece of graphite, is composed of around one trillion trillion carbon atoms. As atoms do not vary that much in size, it is clear that a handful of any solid matter will contain somewhere in the region of a trillion trillion atoms. A trillion trillion is one followed by twenty-four zeros: 1,000,000,000,000,000,000,000,000

It used to be said that it would always be impossible to see pictures of atoms. At least, that is what I was told at school. But this was before 1981, when researchers Gerd Binnig and Heinrich Rohrer at IBM Zurich invented the device known as the scanning tunnelling microscope (STM). The STM maps out the contours of the electric fields around atoms with a resolution of about one-tenth of a nanometre. It is constructed around an extremely fine needle with a tungsten or gold tip that tapers down to a single atom. As the needle scans across a sample, an electronic feedback mechanism controls the position of the needle tip to a precision that is within a fraction of an atomic diameter. The needle scans the surface of the sample at different heights, and these measurements are fed into a computer, where three-dimensional rendering software is used to generate an image of the surface. The result is a remarkable picture that shows the positions of individual atoms on the surface. In Figure 25, each of the tall peaks corresponds to a single atom. Even more amazing is the distribution of the electron waves that can be seen between the atoms. The IBM researchers were stunned by the beautiful images they were creating. 'I could not stop looking at the images,' said Binnig. 'It was like entering a new world. This appeared to me as the unsurpassable high-light of my scientific career and, therefore, in a way, its end.[17]

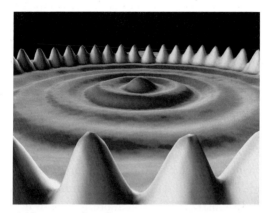

Figure 25 'Quantum Corral', an image produced with a scanning tunnelling microscope. This image represents the electric fields of a ring of forty-eight iron atoms on a copper substrate, and the electron waves confined within this 'corral' of atoms. The iron atoms were individually placed in position by the STM.[18]

The STM can even be used to pick up and arrange individual atoms. This astonishing capability of the machine was used by another IBM employee, Don Eigler, to spell out the name of their company with thirty-five xenon atoms positioned on the surface of a nickel crystal. Binnig and Rohrer were rewarded for their remarkable work with the 1986 Nobel Prize in Physics.

Henry Moseley

Rutherford's discovery of the nucleus was the key to unlocking the structure of the atom. Attention now turned to revealing the secrets of the nucleus itself. In the years just before the First World War, Henry Moseley, a member of Rutherford's team in Manchester, performed a series of experiments proving that the hydrogen nucleus is the basic building block of all atomic nuclei. Later, Rutherford showed that the nucleus of hydrogen is identical to the particle we know as the proton. It is the protons within each nucleus that give the nucleus its electric charge. The nucleus of hydrogen has a charge of +1, because it consists of a single proton; the nucleus of helium has a charge of +2, because it contains two protons; the nucleus of lithium has a charge of +3, as it contains three protons; and so on.

The elements in the periodic table, as originally conceived by Mendeleyev, were ordered by atomic mass. Moseley conclusively demonstrated[19] that a more natural way to order the elements is by their atomic number, which is equal to the charge on the nucleus of the atom, or the number of protons in the nucleus. This number dictates the chemical properties of the atom because it determines the number of electrons in a neutral atom, and it is the electrons that are responsible for the chemical reactions of the atom.[20] Atomic masses steadily increase as the atomic number increases, so Moseley's insight does not actually change Mendeleyev's ordering of the elements, but it does give us a much better understanding of the origin of the periodic table.

Using X-ray spectroscopy,[21] Moseley was able to determine the charge of the nucleus in all the elements between aluminium, which is element number 13, and gold, which is element number 79. Moseley's refined periodic table now showed just four gaps, corresponding to unknown elements with atomic numbers 43, 61, 72 and 75. When his findings were published, the search began for the missing elements. All four are extremely rare, and it was several years before they were discovered. We now know elements 43, 61, 72 and 75 as technetium, promethium, hafnium and rhenium, respectively. Rutherford considered Moseley's achievement to be on a par with the original discovery of the periodic table. The French chemist Georges Urbain went so far as to say that 'The work of Moseley substituted a quite scientific precision for the rather romantic classification of Mendeleyev.'[22]

On the outbreak of war in 1914, Moseley volunteered for service with the Royal Engineers and was commissioned as a signals officer. In 1915, he fought at Gallipoli and on 10 August, during a Turkish attack on the British forces, was shot in the head and killed by a sniper as he was telephoning through an order.[23] Alongside the millions of others who died tragically in this futile conflict, the world lost one of its greatest scientists at the age of just twenty-seven. Had he lived, Moseley would almost certainly have been a recipient of a Nobel Prize and would surely have made many further great contributions to science.

Seeking the First Principles of Things

During the war, Rutherford was appointed to the Royal Navy's Anti-Submarine Division. In parallel with their part in the war effort, Rutherford and his team continued their investigations of the structure of the atom. It was just a few years since his discovery of the nucleus, but now Rutherford was ready to focus his attention on the composition of the nucleus itself. Instead of bombarding high-mass atoms such as gold, he now targeted low-mass atoms, such as nitrogen, believing that by using powerful alpha particle sources he might be able to disrupt their nuclei. And he achieved just that. In about one in three hundred thousand cases, an alpha particle would smash into and merge with a nitrogen nucleus, and a hydrogen nucleus would be spat out. The nucleus that remained was a nucleus of the element oxygen. This process can be represented as follows:

$$\alpha + N \rightarrow O + p$$

Here α represents the alpha particle, which is a helium nucleus containing two protons; N is the nitrogen nucleus, containing seven protons; O is the oxygen nucleus, containing eight protons; and p is a proton, which is equivalent to a hydrogen nucleus. There are thus nine protons before and after the reaction.

Rutherford thus confirmed his suspicion that atomic nuclei really do contain protons, the same fundamental particle that forms the nucleus of the hydrogen atom. And simultaneously he had shown that it is possible to artificially transform one type of atom into another. He had discovered the secret to the age-old dream of the alchemists: the transmutation of the elements. Rutherford first achieved this nuclear transformation in 1917. During a critical point in the research, he was so occupied with his work in the laboratory that he missed one of the Anti-Submarine Division meetings. He telegraphed a message to explain his absence: 'If, as I have reason to believe, I have disintegrated the nucleus of the atom, this is of greater significance than the war!'[24] The response of the Admiralty is not recorded.

In 1919 Rutherford returned to Cambridge to take over from J.J. Thomson as the new director of the Cavendish

Laboratory. The discoveries continued unabated. For atoms other than hydrogen, protons account for only about half the mass of the nucleus. Within a couple of years of his discovery of the proton, Rutherford was already predicting the existence of an uncharged sister particle to the proton. This particle, the neutron, was eventually discovered by James Chadwick working under Rutherford at the Cavendish Laboratory in 1932.

Rutherford had a wonderful intuition for the simplest and most direct ways to investigate natural phenomena, and he coupled this with an unfailing instinct for teasing out what was relevant and important in his experimental results. The research carried out by Rutherford and his teams completely transformed our understanding of the structure of matter. It is difficult to do justice to the magnitude of these discoveries. Although he was a gargantuan figure in the world of physics research, Rutherford lived very modestly. He didn't own his own home until he bought a country cottage in Wiltshire in his later years, and he was never interested in any financial rewards that his research might bring. He typified the traditional British approach to research in which a limited budget was stretched as far as possible and experimenters constructed their own apparatus with great skill and ingenuity. Rutherford and his teams took this style of physics as far as it could possibly go. The next generation of scientists experienced the Second World War and the arrival of the era of Big Science, where serious expenditure was required and physicists had to employ teams of engineers to build their sophisticated equipment.

Rutherford received numerous accolades throughout his long and illustrious career. He was awarded a Nobel Prize in 1908. Surprisingly, it was the Nobel Prize in Chemistry, rather than physics. In 1914 he was knighted. Many other honours followed, and in 1931 he was raised to the peerage as Ernest Lord Rutherford, first Baron Rutherford of Nelson. This entitled him to design a coat of arms for himself (Figure 26). He constructed it with references to his physics research and his native New Zealand. The shield is flanked by a Maori on one side, and the ancient mythical alchemist Hermes Trismegistus on the other. The crest is a baronet's coronet surmounted by a kiwi. The motto beneath the shield is a quotation from the ancient Roman poet Lucretius: *Primordia quaerere rerum*, which means 'To seek the first principles of things'. Lucretius was the author of a remarkable poem, *On the Nature of Things*, which includes the most complete account of the philosophy of atomism to have survived from antiquity.

Figure 26 The coat of arms of Baron Rutherford of Nelson.[25]

The Exclusion Principle

Moseley had demonstrated that the sequence of the elements in the periodic table was determined by atomic number. However, the origin of the periodic patterns in their chemical properties remained unsolved. Bohr had attempted to build on his triumphant explanation of the spectral lines by answering this question, but without complete success. In 1923 he challenged the brilliant young Austrian physicist Wolfgang Pauli to complete the job and produce the first compelling account of the form of Mendeleyev's table based on the structure of atoms. Pauli realised that there was a key ingredient missing from Bohr's recipe for building atoms. He spotted a strange property of electrons that would enable him to succeed. This property is a purely quantum phenomenon that has no counterpart in classical physics, but it is fundamental to the structure of matter. Electrons behave in a very curious way. They are totally antisocial: no two electrons can simultaneously form exactly the same wave. If they did, their waves would cancel each other out. Pauli called this the exclusion principle.

There is an additional feature of electrons that must also be taken into account. Electrons spin at a fixed rate that cannot be changed, although the axis around which the electron spins may change. This means that at most two electrons can occupy the same wave in an atom, but only if their spins are oppositely aligned, with one, say, spinning clockwise and the other anticlockwise. (This being the quantum world, 'spin' as a property is rather more abstract than our everyday notion of a ball rotating on an axis, but that picture is a helpful analogy.[26])

If it were not for the exclusion principle, all electrons would sink down to the lowest rung of the energy ladder. There would then be no chemical bonding between different atoms, and so none of the cornucopia of different molecules that produce the diverse materials of our world could exist. Another look at our imaginary one-dimensional, guitar-string atoms will illustrate the effect of the exclusion principle. Imagine adding electrons one at a time. The first electron drops down the rungs of the energy ladder until it reaches the fundamental mode of vibration. It cannot fall any further in energy. In a hypothetical world of guitar-string atoms, this would be equivalent to element number 1, hydrogen. A second electron would also fall down to the lowest energy rung. It can occupy this energy level if its spin is opposite to that of the first electron. We can imagine one electron spinning clockwise around the line and the other spinning anticlockwise. This is the guitar-string atom equivalent of element number 2, helium. The lowest rung on the energy ladder contains two electrons, with opposite spins. No more electrons can exist at this rung of the energy ladder, so if we add a third electron to the atom it will fall only as far as the second energy rung. As we add more electrons, they will fill up the energy rungs in ascending order, with no more than two on each, as shown in Figure 28 (see page 140).

The Atomic Drum

In our previous model the electron was allowed to move only in one dimension, and the electron's wave was analogous to the vibrations of a guitar string. If we add another dimension to the model, and let the electron inhabit a square, two-dimensional region, we will be able to see why symmetry is so important in quantum mechanics. The waves formed by the electron in our new model are similar to the vibrations of a square drum.[27] The lowest-energy vibration of the square drum is one in which the whole drum skin vibrates in and out together. This wave is the fundamental mode of vibration of the drum and occupies the lowest rung on the drum's energy ladder, but the symmetry of the square drum produces a new feature at the next rung on the ladder. In one mode of vibration, the left and right halves of the drum alternately oscillate in and out. (A horizontal cross section of this vibration would appear like the second harmonic of the vibration on a string.) But there is also a mode in which the top and bottom halves of the drum alternately oscillate in and out. This is a different mode of vibration, but the energy of the two modes must be exactly the same.

A square is symmetrical under a 90-degree rotation, which means that if we rotate the square drum through 90 degrees it will appear unchanged. (Alternatively, we could leave the square where it is and move ourselves 90 degrees around it and it will appear exactly the same.) Although the square itself is unchanged, the two modes of vibration that we have been considering have now been swapped over. The energy of the two modes must be identical, because there is no way to tell them apart; the only difference between them is in our orientation with respect to the square. So on the second rung of the energy ladder there are two different waves available to the electron which have exactly the same energy. As we move up the energy ladder, most of the modes of vibration of the drum will come in such pairs.[28]

The lowest modes of vibration of the square drum are depicted in Figure 27. The portion of the drum that is depressed into the page is shown shaded and the portion of the drum that is raised out of the page is white. A vibration is represented as two squares connected by a double-headed arrow. Each square shows a momentary snapshot of the vibration, but at the opposite point in its cycle to the other square in the pair. Each vibration is an oscillation between these two positions. In the figure, the energy of the modes of vibration increases from bottom to top. At the first and third rungs on the energy ladder there is just one mode of vibration. But at the second and fourth rungs there are two different modes of vibration with the same energy.

Chemists refer to a collection of electrons in an atom that have the same energy as a shell. As each wave contains, at

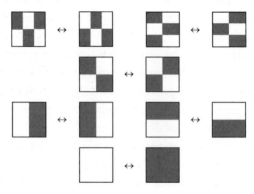

Figure 27 Waves on a square drum (or electron waves in a hypothetical square atom). The energy of the waves increases from bottom to top.

most, two electrons with opposite spins, in our guitar-string atoms there are at most two electrons in each shell. In our square-drum atoms the symmetry of the square means that some of the waves are matched with other waves with exactly the same energy. Taking into account the two spin states, this means that some of the shells of these atoms can contain up to four electrons. The number of electrons in each shell as we ascend the energy ladder of the square-drum atoms goes in the following sequence: 2, 4, 2, 4, 4, 2, … . Each electron falls into the lowest-energy wave that is available, so the shells are progressively filled up, and the electrons fill each shell completely before any of the waves in the next shell above are occupied.

Atoms bond with other atoms to form molecules. This is what chemistry is all about: the shuffling of atoms between molecules. Chemical reactions are the processes in which bonds between atoms form and break. It is only the electrons in the uppermost, partially filled shells that participate in the chemical bonding between atoms. The number of electrons in the uppermost shell determines the chemical properties of the atom, such as the number of bonds the atom can form with other atoms – what chemists call valency. Two atoms with the same configuration of electrons in this upper shell will undergo similar chemical reactions and have similar properties, and in the periodic table they will be located in the same column.

If atoms really were one-dimensional, like guitar strings, the periodic table would be very boring. There would be only two types of atom. Atoms with an odd number of electrons would have a single electron in their uppermost shell, while

---

Puzzle 5

Think about the relationship between symmetry and the existence of different waves with the same energy. What happens if we remove the symmetry by stretching our hypothetical square atom to form a rectangle? If the atoms were rectangular rather than square, they would no longer be symmetric under 90-degree rotations. Why would this loss of symmetry affect the equality of the energy levels of the waves?

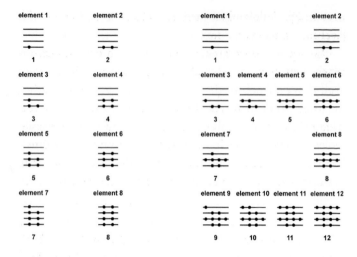

Figure 28 Left: How the periodic table would look if atoms were one-dimensional, showing schematic representations of the energy levels of the electrons in the atoms of each element. Right: The structure of the periodic table if atoms were square.

atoms with an even number of electrons would have two electrons in their uppermost shell. We can imagine that all the odd elements would share similar chemical properties to one another, as would all the even elements. Such a string orchestra periodic table is shown on the left in Figure 28. Each line represents an energy level in the atom and each dot on a line represents an electron that is situated at that energy level.

If atoms were square, the periodic table would be more interesting as there would be elements whose atoms have 1, 2, 3 or 4 electrons in their uppermost shell. The square-drum periodic table is shown on the right in Figure 28. If the elements of our imaginary one-dimensional world are like a string orchestra, then the elements of our two-dimensional world might be like the mellifluous clanging percussion of a Javanese gamelan ensemble.

Our square-drum atoms have given us a first indication of why symmetry plays such an important role in a quantum mechanical description of the universe. The wave-like nature of electrons makes them sensitive to the shape of the region they occupy. In classical physics, particles are considered to behave like billiard balls. A billiard ball will roll around the billiard table and bounce off the cushions, but it is only aware of the particular point on the table that it is

---

Answer to Puzzle 5

By stretching our square atoms into rectangles, we destroy the symmetry of the atoms under 90-degree rotations. We would then have no reason to expect the energy of waves that are swapped over by such a rotation to be the same. If we consider the first two waves that form such a pairing in the square atoms, we can see that they do not have the same energy if the atoms are rectangular. These two modes of vibration are shown below. The two sides of the rectangle have different lengths, so the vibrations in the vertical direction have a different wavelength from the vibrations in the horizontal direction. We know that shorter wavelength corresponds to higher energy, so the wave on the left in the diagram has lower energy than the wave that is shown on the right.

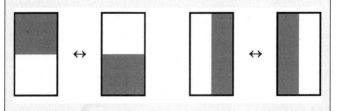

---

touching at any one time: it is completely oblivious to the overall shape of the table. However, a real subatomic particle, such as an electron, behaves in a very different way, a quantum way. An electron is associated with a wave: the wave spreads out to fill the region in which the electron is contained, and the form the wave takes depends on the shape of this region. This is certainly very strange, but it also appears to be very true.

The Real Periodic Table

Bohr's model of the atom was a great leap forward, but it was not a secure foundation for a totally new theory of matter, as it was based on what seemed to be some quite arbitrary assumptions. The interpretation of the model described above is somewhat anachronistic, as Bohr did not originally consider his model in terms of electron waves, but the model becomes clearer with the benefit of this insight. It was not until the middle years of the 1920s that the fundamental principles of quantum mechanics were gradually elucidated, and a mature formulation of quantum mechanics arrived. This new understanding developed out of a proposal by a young aristocratic French physicist called Louis de Broglie (pronounced 'de Broy'). It was de Broglie who suggested, in his PhD thesis, that just as light, which was traditionally thought of as a wave, is composed of particles, so the particles that form matter, such as electrons, might display wave-like properties. This idea transformed physics. The components of matter were now recognised as having a dual nature. According to de Broglie's quantum hypothesis, they are able to simultaneously behave like both waves and particles.

De Broglie's proposal was almost immediately developed into two new and complementary approaches to quantum mechanics. In 1925 a young German theorist, Werner Heisenberg, devised a rather abstract technique that he called matrix mechanics. The following year an Austrian physicist, Erwin Schrödinger, using more familiar methods, developed an alternative route to quantum mechanics that became known as wave mechanics. This was based on solving an equation that describes the form that is taken by matter waves such as electron waves. It is called the Schrödinger equation, and is one of the most important equations ever devised. As soon as Schrödinger applied his equation to the hydrogen atom, he could see that he had discovered something very special. The solutions gave all the same energy levels that Bohr had found, which was good news as this was the great success of the Bohr model. But Bohr's methods were rather ad hoc: it was difficult to see how they could be applied to other problems. In contrast, it was clear that by following Schrödinger's approach to quantum mechanics, a Schrödinger equation could be formulated for many different types of physical problem. By solving this equation it would be possible for the first time to investigate many esoteric areas of fundamental physics. The clarity of Schrödinger's technique would also help to provide a better understanding of the true nature of quantum mechanics.

Solving the Schrödinger equation gives physicists a coherent picture of the electron waves in an atom. The great disparity in size between an atom and its nucleus might suggest that most of the atom is nothing but empty space. However, this is not a very accurate image. The Schrödinger equation implies that the electrons in an atom form three-dimensional waves that spread throughout the space occupied by the atom. In this picture, the electrons are often described as forming a cloud around the nucleus. This makes them much more difficult to imagine than the waves on a guitar string or on a square drum, but the result is that atoms are 'solid' in the sense that if we try to squeeze them, we find they are difficult to compress. Squeezing an atom will confine its electron cloud into a smaller space, which will reduce the wavelengths of the electrons and thus raise their energy. Consequently, the electrons resist being confined into a smaller space, and a lot of energy must be applied as the electrons' energy increases and the atom is compressed.

The Schrödinger equation for real atoms has a much higher degree of symmetry than does our square-drum model. Whereas the symmetry group of our two-dimensional square-drum atom is the symmetry group of a square, the symmetry group of real three-dimensional atoms is much larger, so there are more distinct waves at each energy level than for the imaginary square-drum atom. Each shell in an atom consists of twice the number of the waves that are available to electrons at a particular energy level. (This is because the spin of each electron can be oriented in either of two opposite directions.) Using the Schrödinger equation, we can find all the possible electron waves in real atoms and double this number to work out the number of electrons that will fit in each shell. We find that the number of electrons that will fit into the shells in real atoms form the following sequence in ascending energy: 2, 8, 8, 18, 18, 32, 32, ...[29] These numbers give us the number of elements in successive rows of the periodic table. This can be seen most clearly in the 'long-form' representation of the periodic table, as shown in Figure 29. The bottom row of 32 elements is incomplete, as the synthesis of highly radioactive artificial elements beyond number

Figure 29 The 'long-form' periodic table. The numbers down the left-hand side are the numbers of electrons in successive atomic shells, and therefore also the number of elements in each of the rows, or 'periods', of the table.

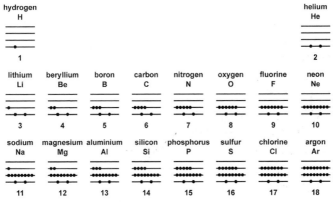

Figure 30 Atomic structure and the periodic table: a schematic representation of the energy levels in atoms. The lines represent the energy levels in successive shells; the dots represent the electrons in the shells.

112 has not yet been confirmed in the laboratory. Figure 30 is a schematic representation of the number of electrons in each shell for the elements in the first three rows of the periodic table.

The arrangement of the electrons in atoms is closely related to the chemical properties of the elements. The most obvious example is the column on the extreme right of the periodic table, which contains the inert gases:

> ### Puzzle 6
> How are the electrons arranged in element number 19, potassium?

helium, neon, argon, krypton, xenon and radon. The electrons in the atoms in each of these gases form completely filled shells, so they have no electrons in partially filled shells where they would be available to form bonds with other atoms. This is why they are so unreactive and rarely form molecules. By contrast, the column on the extreme left of the table contains the alkali metals, which are highly reactive. They each have a single electron in their outermost shell. This electron is only loosely bound in the atom and is available for bonding with other atoms.

The pioneers of quantum theory produced a triumphant explanation of the origin of Mendeleyev's periodicity of the elements. But this was just the start.

It's a Crazy Old World

Matrix mechanics and wave mechanics, the two distinct approaches to quantum theory, relied on very different mathematical techniques, but were soon shown to be completely equivalent. Theorists now had the mathematical means to tackle a wide range of physical problems, and over the following decade that is exactly what they did. Quantum mechanics was applied everywhere. Whole new branches of physics were born as many of the properties of matter could now be explained for the first time. To give just a few examples: it became possible to understand how atoms bond together to form molecules and how chemical reactions occur; the properties of crystals and other solids could be investigated at the atomic level; the structure of the atomic nucleus and the origin of radioactivity became explicable; and, as we will see in a later chapter, it became possible to explain how stars are able to shine for billions of years.

Much modern technology depends on the principles of quantum mechanics for its operation, and we are set to become ever more dependent on gadgets that rely on mysterious quantum phenomena. The twenty-first century will be the century of quantum technology. Our understanding of physical effects which are essentially quantum mechanical in nature is critical in the technology behind the semiconductors from which the memory and processor chips in our computers are fabricated, the lasers that read our DVDs, the holograms on our credit cards, solar cells, digital cameras, LCD screens, electron microscopes, and superconducting magnets in MRI scanners. The utilisation of quantum mechanics is promising to open up areas such as nanotechnology, quantum cryptography and quantum computing. But the importance of quantum mechanics goes far beyond even these examples. The fundamental structure of matter, and consequently the entire science of chemistry, cannot be understood without quantum mechanics. In fact, as we now know, the very existence of atoms depends on quantum mechanical effects, because quantum mechanics applies to all elementary particles and how they interact. Our understanding of the laws of the universe, and the whole of modern physics, is built on quantum mechanics. To a modern physicist, a world without quantum mechanics is inconceivable.

The quantum revolution implies that, although in many respects light behaves like a wave, it also has particle-like characteristics and is composed of particles that we call

---

Answer to Puzzle 6

There are nineteen electrons in a neutral potassium atom (i.e. one that carries no electric charge as it contains as many electrons as there are protons in its nucleus). Eighteen of these electrons fill the first three shells, giving an electronic structure like an argon atom. As the third shell is now full, the final electron must go into the fourth shell.

---

photons. Conversely, the elementary particles from which matter is composed, such as electrons and protons, also show wave-like characteristics. Both matter and radiation exhibit a curiously ambiguous wave–particle duality. Despite the great successes of quantum mechanics, there is no disguising the fact that there is something very weird about the idea, that the same entity is sometimes best described as a wave and sometimes as a particle. The deeper we delve into the philosophical implications of quantum mechanics, the weirder it gets. Niels Bohr used to say that anybody who wasn't shocked by quantum mechanics didn't understand it. But the flamboyant Richard Feynman, one of the heroes of our next chapter, went even further, and would point out that nobody really understood quantum mechanics. This is how he presented this startling admission in a popular exposition:

> What I am going to tell you about is what we teach our physics students in the third or fourth year of graduate school – and you think I'm going to explain it to you so you can understand it? No, you're not going to be able to understand it … You see, my physics students don't understand it either. That is because I don't understand it. No-body does.[30]

And this is how he introduced the subject to his students:

> Because atomic behaviour is so unlike ordinary experience,
> it is very difficult to get used to it and it appears peculiar and
> mysterious to everyone, both to the novice and the experi-
> enced physicist. Even the experts do not understand it the
> way that they would like to, and it is perfectly reasonable
> that they should not, because all of direct, human experience
> and of human intuition applies to large objects. We know
> how large objects will act, but things on a small scale just do
> not act that way. So we have to learn about them in a sort of
> abstract or imaginative fashion and not by connection with
> our direct experience.[31]

Our intuitive understanding of reality is derived from
our everyday interactions with the world around us. We
have a concept of waves from seeing waves on a pond, or
a vibrating guitar string. We have a concept of particles
from playing snooker or pool. When we think about atoms
and the other constituents of matter, it is natural to try
to make use of concepts we are familiar with. But these
concepts seem to be no longer applicable when considering
the behaviour of atoms and electrons and photons. Photons and electrons are really neither waves nor particles,
and we do not have an everyday term that would adequately
describe their behaviour.

Fortunately, the mathematics of quantum mechanics works extremely well, whatever we may think about
the crazy world it is describing. It is quite marvellous that,
although it is so difficult to conceive of what goes on in the
subatomic domain, we can still apply the rules of quantum
mechanics to calculate the values of physical quantities

with great accuracy. Unintuitive it may be, but, as we will
see in the next chapter, the predictions of quantum physics have been thoroughly tested in the laboratory – and
have been found to agree with the experimental results
to a mind-boggling precision. Quantum theory has so
far passed every test it has been subjected to, without
a single exception.

Chapter Four

QED

To Brocken's tip the witches stream,
The stubble's yellow, the seed is green.
There the crowd of us will meet.
Lord Urian has the highest seat.

Goethe, Witches' chorus in *Faust* (1808)[1]

Scotch Mist

In 1894, the physicist C.T.R. Wilson spent a few weeks away from the Cavendish laboratory in Cambridge working as a meteorological observer on the summit of Scotland's highest mountain, Ben Nevis. For much of his time there, mist and cloud shrouded the mountain tops around the research station. One day, while Wilson was making observations up in the mist, he was confronted by a huge ghostly figure surrounded by a multicoloured aura. This was the Brocken spectre, the giant who would direct Wilson towards his future research.

If weather conditions are just right, hikers in the mountains can witness the phenomenon of the Brocken spectre (see Plate 5). When the Sun hangs low in the sky and shines through the misty gloom, it may cast a shadow onto a bank of fog. Hikers facing away from the Sun will sometimes see a giant shadow of themselves enveloped in an aetherial multicoloured halo. The coloured rings forming the halo are optical effects known as glories. They are produced by sunlight that is refracted and scattered back by uniformly sized water droplets and appear as circular rainbows. Any movement of the mist enhances the eerie appearance of the giant figure by making the looming spectre seem to hover in the air. The Brocken spectre takes its name from the highest peak of Germany's Harz Mountains, a peak that often projects through the cloud layer and provides a good display of the giant. This was the ominous site chosen by Goethe for the scene of the Witches' Sabbath on Walpurgis Night in his epic drama, *Faust*. Goethe refers to the giant as Herr Urian – Lord Urian in the above translation. Wilson may not have sold his soul, as Faust did, but his experience of the spectre on the summit of Ben Nevis would lead him to the highest accolade for a physicist – a Nobel Prize.

Charles Thomson Rees Wilson (1869–1959) was born into a long line of Scottish farmers in the village of Glencorse in Midlothian. His father died when he was aged just four, and his mother took the family to live in Manchester. He grew up in this city and studied biology at Owen's College (which later became the University of Manchester) with the intention of training to be a doctor. He won a scholarship to Cambridge University, and Sidney Sussex College became his home for the rest of his career.

Wilson was intrigued by the spectacular optical phenomena he had witnessed in the mountains. When he returned to Cambridge early in 1895, he set about reproducing the optical effects he had seen in Scotland. To do this he needed a reliable laboratory method of creating miniature clouds that resembled the real thing, so he built a sealed chamber containing air that was saturated with water vapour. He then attached a piston to rapidly expand and cool the air inside the chamber. As humid air cools, its ability to retain water vapour is reduced, and so the water vapour is brought to the brink of condensation. But water droplets can form only if there are dust particles or tiny ice crystals present, around which the water can condense. Wilson soon discovered, however, that even when the vapour was completely dust free, a few small water droplets would always form in his chamber.

After a couple of months of studying cloud formation in his device, Wilson speculated that these droplets might be forming around charged ions in the vapour. The newly discovered X-rays were known to ionise air, so he tested his idea by firing X-rays into the cloud chamber. His suspicions were confirmed. The path of an X-ray through the chamber was made visible as a vapour trail of tiny water droplets. What was happening was that when a high-energy X-ray photon or a charged particle passed through the cloud chamber, it collided with electrons in atoms in the vapour. The electrons were knocked out of the atoms, leaving charged ions behind. Water vapour then condenses around the ions, and the path of the particle becomes visible as a vapour trail. Wilson realised that he had developed the perfect instrument for detecting the presence of any electrically charged fundamental particles.

Over the course of many years, Wilson perfected his cloud chamber. By placing the chamber in a magnetic field he was able to determine the electric charge of a particle passing through it. In a magnetic field, charged particles follow curved trajectories, with positive and negative particles curving in opposite directions.[2] The cloud chamber was instrumental in the first discovery of a subatomic particle – the electron. In a series of methodical experiments in the late 1890s, Wilson's boss J.J. Thomson, the director of the Cavendish Laboratory in Cambridge, gradually elucidated the properties of the electron.[3] The decisive factor in gaining Thomson the credit for the discovery of the electron was his access to a crucial piece of equipment, Wilson's cloud chamber, which he used to measure the electric charge of the electron. Thomson was awarded the 1906 Nobel Prize in Physics for this discovery. Wilson would receive the same prize twenty-one years later for his invention of the cloud chamber, one of the most important tools of twentieth-century physics. The cloud chamber would play a decisive role in the career of another Cambridge physicist – the theorist Paul Dirac.

P.A.M. Dirac

Paul Adrien Maurice Dirac was one of the greatest theoretical physicists of the twentieth century. Born in Bristol in 1902, he was the second son of Charles Dirac, a French-language tutor and an immigrant from Switzerland. His mother was the daughter of a Cornish sailor. Paul had an elder brother,

Félix, and a younger sister, Béatrice. The household was not a happy one: the family suffered at the hands of Dirac's domineering father, who insisted that all conversation at the dinner table should be in French:

> My father made the rule that I should only talk to him in French. He thought that it would be good for me to learn French in that way. Since I found that I couldn't express myself in French, it was better for me to stay silent than to talk in English. So I became very silent at that time – that started very early.

Throughout his life he would have a reputation for the rarity of his utterances. The style of Dirac's mathematics and writing displays the same directness as his speech, being clearly thought through, concise and straight to the point:

> My father always encouraged me toward mathematics ... He did not appreciate the need for social contacts. The result was that I didn't speak to anybody unless spoken to. I was very much an introvert, and I spent my time thinking about problems in nature.[4]

His brother Félix committed suicide in 1925. Dirac blamed their severe father, and completely dissociated himself from him after this tragic event.

Dirac completed a degree in electrical engineering at the University of Bristol and remained at the university studying applied mathematics. In 1923 he was awarded a research scholarship to study at St John's College, Cambridge, and it was there that he produced the work on fundamental physics for which he is famous. His early years in Cambridge coincided with the quantum mechanical discoveries of Louis de Broglie, Werner Heisenberg and Erwin Schrödinger, that had thrown physics into turmoil. Each of these breakthroughs was of the utmost significance, and each would be rewarded with a Nobel Prize, but it wasn't yet clear how all the new ideas fitted together. Dirac set about investigating the new physics and put his mind to finding the best way to formulate the mathematics of quantum mechanics. He was a very precise thinker who was guided by symmetry and elegant mathematics. He succeeded in reorganising the foundations of quantum mechanics so that it could be expressed with a clarity that would prove vital for further theoretical developments. Most physicists would agree with Einstein, who said that in his opinion it was to Dirac that 'we owe the most logically perfect presentation' of quantum mechanics.[5] In 1930, Dirac published his textbook *The Principles of Quantum Mechanics*. Its importance can be gauged by its longevity in a subject that thrives on fresh ideas: it is still in print after more than eighty years.

In 1929, Dirac was invited to lecture at the University of Wisconsin and the University of Michigan. During his stay in Madison, the *Wisconsin State Journal* published an interview with him[6]:

ROUNDY INTERVIEWS PROFESSOR DIRAC
AN ENJOYABLE TIME IS HAD BY ALL

I been hearing about a fellow they have up at the U. this spring – a mathematical physicist, or something, they call him – who is pushing Sir Isaac Newton, Einstein and all the others off the front page. So I thought I better go up and interview him for the benefit of *State Journal* readers, same as I do all other top notchers. His name is Dirac and he is an Englishman. He has been giving lectures for the intelligentsia of the math and physics departments – and a few other guys who got in by mistake.

So the other afternoon I knocks at the door of Dr. Dirac's office in Stirling Hall and a pleasant voice says 'Come in.' And I want to say here and now that this sentence 'come in' was about the longest one emitted by the doctor during our interview. He sure is all for efficiency in conversation. It suits me. I hate a talkative guy.

I found the doctor a tall youngish-looking man, and the minute I seen the twinkle in his eye I knew I was going to like him. His friends at the U. say he is a real fellow too and good company on a hike – if you can keep him in sight, that is.

The thing that hit me in the eye about him was that he did not seem to be at all busy. Why if I went to interview an American scientist of his class – supposing I could find one – I would have to stick around an hour first. Then he would blow in carrying a big briefcase, and while we talked he would be pulling lecture notes, proof, reprints, books, manuscript, or what have you out of his bag. But Dirac is different. He seems to have all the time there is in the world and his heaviest work is looking out the window. If he is a typical Englishman it's me for England on my next vacation! Then we sat down and the interview began.

'Professor,' says I, 'I notice you have quite a few letters in front of your last name. Do they stand for anything in particular?'

'No,' says he.

'You mean I can write my own ticket?' 'Yes,' says he.

'Will it be all right if I say that P.A.M. stands for Poincaré Aloysius Mussolini?'

'Yes,' says he.

'Fine,' says I, 'We are getting along great! Now doctor will you give me in a few words the low-down on all your investigations?'

'No,' says he.

'Good,' says I. 'Will it be all right if I put it this way – "Professor Dirac solves all the problems of mathematical physics, but is unable to find a better way of figuring out Babe Ruth's batting average"?'

'Yes,' says he.

'What do you like best in America?', says I. 'Potatoes,' says he.

'Same here,' says I, 'What is your favourite sport?' 'Chinese chess,' says he.

That knocked me cold! It was sure a new one on me! Then I went on: 'Do you go to the movies?'

'Yes,' says he. 'When?' says I.

'In 1920 – perhaps also 1930,' says he.

'Do you like to read the Sunday comics?'

'Yes,' says he, warming up a bit more than usual.

'This is the most important thing yet, doctor,' says I. 'It shows that me and you are more alike than I thought. And now I want to ask you something more: They tell me that you and Einstein are the only two sure-enough high-brows and the only ones who can really understand each other. I won't ask you if this is straight stuff for I know you are too modest to admit it. But I want to know this – Do you ever run across a fellow that even you cant understand?'

'Yes,' says he.

'This will make a great reading for the boys down at the office,' says I. 'Do you mind releasing to me who he is?'

'Weyl,' says he.[7]

The interview came to a sudden end just then, for the doctor pulled out his watch and I dodged and jumped for the door. But he let loose a smile as we parted and I knew that all the time he had been talking to me he was solving some problem that no one else could touch.

But if that fellow Professor Weyl ever lectures in this town again I sure am going to take a try at understanding him! A fellow ought to test his intelligence once in a while.

Antimatter

Schrödinger had come up with an equation that could be used to describe electrons in atoms, but it did not take into account the other great revolution of early twentieth-century physics – relativity. Einstein's theory implied that although Newtonian mechanics works extremely well, its predictions fail for objects moving at speeds approaching the speed of light. This is because the speed of light is the universe's ultimate speed limit: it is the maximum speed allowed by the laws of nature.[8] It also places a limit on the time it takes for objects to interact with one another. For instance, the electro-magnetic force between two charged particles will only be felt after an interval sufficient for the influence to have travelled between the two particles at the speed of light. This information is automatically encoded in the equations that Maxwell deduced from Faraday's experimental results. It is quite remarkable that Maxwell's equations, which encapsulate the behaviour of charged particles and electromagnetic waves, are fully compatible with relativity, even though they were derived by Maxwell fifty years before Einstein's first papers on relativity.[9] In fact, it was the need to develop a theory of mechanics consistent with Maxwell's equations that provided Einstein with the motivation to develop his theory of special relativity.

Dirac knew that the next logical step in the development of quantum mechanics was to formulate the theory in a way that was consistent with relativity. If quantum theory was going to be applicable to particles travelling at speeds approaching the speed of light, then it would have to agree with relativity. The electromagnetic force is transmitted at the speed of light, so relativity would necessarily be a vital component of a quantum theory of electromagnetism. Dirac always strove for mathematical elegance. He felt intuitively that the correct description of fundamental physics had to be simple and uncluttered, and his mathematics always possessed an austere beauty. Symmetry lay at its heart. When Dirac was asked to summarise his philosophy of physics, he wrote a single sentence on the blackboard[10]:

PHYSICAL LAWS SHOULD HAVE MATHEMATICAL BEAUTY

One night in 1928, Dirac was sitting in his room in St John's College, staring into the fire. As he let his mind drift, watching the flames, he was struck by an idea that would allow him to construct an equation of a completely novel kind.[11] He could see at once that his new approach was going to work. The result was the equation that is now known as the Dirac equation. It describes the behaviour of matter waves, such as electron waves, even when their speed approaches the speed of light, and it was Dirac's greatest achievement. The Dirac equation is one of those instances, like Maxwell's equations, when suddenly everything falls into place and the equations seem to convey more information than was put into them. This was one of those rare occasions when a theoretical physicist immediately knew intuitively that the mathematics must represent a fundamental truth about the structure of the universe.

Dirac's penetrating analysis of his equation persuaded him that there was a previously unsuspected symmetry at the heart of matter. He could not make complete sense of his equation unless the electron had a mirror-image particle, a particle whose mass was the same as the mass of the electron but whose electric charge was positive instead of negative. At the time, the only elementary particles that were known were the electron, the proton and the photon. Most physicists felt that this small collection of particles was sufficient to explain the structure of matter entirely, so the prediction of another fundamental particle would be a very bold move indeed. Despite the huge difference in mass between the proton and the electron, Dirac initially thought that the mirror particle implied by his equation must be the proton. But he soon came to the conclusion that this could not work, so he was forced to postulate the existence of a completely new particle. In 1931 he referred to this particle as the anti-electron.

The Positron

Our planet is continually bombarded by intense radiation from the depths of space. Energetic particles known collectively as cosmic rays rain down on us from the heavens, but fortunately the Earth's atmosphere shelters us from much of this radiation by absorbing it. The cosmic origin of the radiation was conclusively demonstrated in 1912 when an intrepid Austrian physicist, Victor Hess, and two colleagues began a series of balloon flights that reached altitudes of up to 5,000 metres. Their measurements showed that the intensity of the radiation increases with height, so they could definitively rule out a terrestrial origin for it.[12]

An American physicist, Carl Anderson, decided to investigate the nature of cosmic rays after he completed his doctorate in 1930. Although the existence of cosmic rays had been known for almost two decades, their identity was still a mystery. Anderson built a cloud chamber with a higher magnetic field than any previously constructed and started photographing the vapour trails formed by high-energy cosmic rays racing through it. Some of the trails presented him with a puzzle. These paths were curved, as would be expected for charged particles passing through the magnetic field in the chamber. But if they were formed by particles travelling downwards, which they would be if they were coming from outer space, their trajectories were curving in the opposite direction to those formed by electrons. This implied that their charge must be positive. The trails looked as though they had been made by particles that were very similar to electrons, but with opposite electric charge. No such particles were known. Furthermore, the particles could not be protons, because collisions with the atoms in the cloud chamber would rapidly bring a particle as massive as a proton to a halt.

Anderson had to find a way to rule out the possibility that these particles were just ordinary negatively charged electrons travelling upwards, however unlikely that might seem. He had a brilliant idea: he could make a simple modification to the cloud chamber that would enable him to conclusively identify the charge of the particles. By placing a lead plate across the centre of the cloud chamber, he could unambiguously determine the direction in which the particles were travelling. When a particle passed through the lead plate it would slow down. The magnetic field in the cloud chamber would have the greatest effect on the paths of particles which were travelling slowly. So the path of a particle would curve more sharply after it had passed through the lead barrier than it did before it reached the lead barrier. This would be clear in a photograph of the trail of the particle, and would show from which direction the particle had arrived. The modified cloud chamber enabled Anderson to prove that the particles were definitely travelling downwards: the particles were indeed positively charged cosmic rays with a mass similar to that of the electron. *Science News Letter* published the first picture of one of these tracks on 19 December 1931. At the behest of the journal's editor, Anderson reluctantly adopted the name 'positron' for the new particle.[13] 'Positron' is a portmanteau word, as Lewis Carroll would call it, a contraction of 'positive electron'. But it is the name that we still use for these particles today.

In Anderson's photograph, shown in Figure 31, a positron comes into the cloud chamber from above, its path curved by

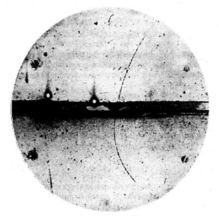

Figure 31 Photograph of the trail of a positron through Carl Anderson's cloud chamber.[14]

the magnetic field. Initially it is travelling very fast, so the curvature in its trajectory is slight. It then passes through the lead plate, which slows it significantly, so its path below the lead plate therefore shows a greater curvature. It was clear that Anderson's positrons were the anti-electrons whose existence was demanded by Dirac's theory. Both Dirac and Anderson were to win the Nobel Prize in Physics. Dirac won his in 1933, sharing the award with Erwin Schrödinger, and Carl Anderson was rewarded for the discovery of the positron with the 1936 prize.

The Dirac equation encapsulates the behaviour of all the different types of particle from which matter is formed. It predicts that there must be a mirror-image particle for other particles, such as the proton and the neutron. This has, indeed, proved to be the case. The antiproton was found in 1955, followed by the antineutron in 1956. These antiparticles will combine to form antimatter in the same way that ordinary particles combine to form matter. For example, a positron will bind to an antiproton to form an atom of antihydrogen, as has now been demonstrated in the laboratory. A facility known as the Antiproton Decelerator has been built at CERN specifically for the study of antihydrogen.[15] When produced in high-energy collisions in an accelerator, antiprotons race off at close to the speed of light. To produce antihydrogen they must be collected and slowed down to a few millionths of this velocity and then mixed with positrons so that they can combine into atoms of antihydrogen. The experiments at CERN, which started in 2002, have already produced tens of thousands of antihydrogen atoms. The aim of these investigations is to measure the energy levels of the positron in an antihydrogen atom to see whether they are exactly the same as the energy levels of the electron in a hydrogen atom, as would be expected if there is an exact symmetry between matter and antimatter. If any slight difference is observed, it might shed light on the mystery of why the universe is composed almost entirely of matter, with only the tiniest residue of antimatter. This critical feature of the universe remains a puzzle for physicists and has still not been fully explained.

The importance of Dirac's insight can be gauged from Heisenberg's remark that Dirac's 'discovery of antimatter was perhaps the biggest jump of all the big jumps in physics of our century. It was a discovery of the utmost importance because it changed our whole picture of matter.'[16]

Mass and Energy

The confirmation of Dirac's interpretation of his equation offered theoretical physicists a deep insight into electromagnetic interactions at the level of fundamental particles. According to Dirac, when an electron and a positron meet they can mutually annihilate and disappear in a burst of high-energy gamma-ray photons. Conversely, under the right conditions it would be possible to convert a gamma-ray photon into a pair of particles, an electron and a positron. This was a completely new idea. Matter was not eternal: it could be created or destroyed. What had driven Dirac towards his equation was his desire to make quantum mechanics compatible with Einstein's relativity. The interconvertibility of matter and radiation that Dirac had deduced from his equation now gave a new significance to the most famous equation that Einstein had derived from his theory:

$$E = mc^2$$

This equation is the best known in the whole of physics. It sums up the equivalence of mass and energy in a statement that could be used as a slogan on a tee-shirt. In the equation E represents energy, m represents mass and c is the speed of light. Light travels extremely fast, so the speed of light is a large number, and therefore its square, c^2, is huge. To find out how much energy is contained in a lump of matter, we must multiply its mass by c^2, so mass is equivalent to an enormous amount of energy. By combining quantum mechanics and relativity, Dirac had shown how the conversion of mass into energy and energy into mass works at the level of particle interactions.

Quantising the Electromagnetic Field

When Faraday first drew his maps of electric and magnetic fields, some nineteenth-century physicists criticised him for filling space with imaginary lines of force. But Maxwell demonstrated the value of Faraday's insight when he used the concept of fields to derive his elegant and highly successful equations of electromagnetism. However, the physical significance of the fields was still not completely clear. Were they just mathematical devices, or did they correspond to some fundamental feature of reality? Did they have any real physical existence, or were they just abstract notions that help us with our calculations? The answer is quite remarkable, but it only became clear with the development of a quantum theory of electromagnetism.

The approach to quantum mechanics developed by Schrödinger in the 1920s is extremely valuable, but it is really only a halfway house to a fully quantum theory. One of Schrödinger's groundbreaking papers analysed the electron waves in the hydrogen atom. The electromagnetic force between the positively charged proton and the negatively charged electron binds the two particles together to form a hydrogen atom. Schrödinger applied his equation to this problem and successfully calculated the energy levels that the electron could occupy. However, although the electron in the hydrogen atom was treated as a wave, the electromagnetic force between the electron and the proton was treated just as it would be in a classical, non-quantum theory. The Schrödinger equation encapsulated the quantum nature of particles such as electrons from which matter is formed, but not the quantum nature of the forces between these particles. A more complete quantum theory would have to treat both matter and forces in the same quantised way. It was clear that to make further progress the particle-like behaviour of the electromagnetic field would have to be incorporated into the theory. After all, the theory of quantum mechanics had originally been launched when Planck and Einstein had realised that electromagnetic waves have a particle-like nature. Light, and indeed electromagnetic radiation of all wavelengths, is composed of particles that we know as photons.

By considering the composition of the electromagnetic field in terms of photons, we can get a picture of how the electromagnetic force works at the level of particle interactions. In the pictures drawn by Michael Faraday, an electric field surrounds a charged particle such as an electron. This is just a map of the electric force around the electron. In the quantum picture, the nature of the electric field becomes much clearer. The field is no longer considered to be continuous throughout space; instead, it is composed of photons – particles of electromagnetic radiation – that flicker in and out of existence. The electron is bathed in a fluctuating cloud of photons that are continually being emitted and re-absorbed by the electron. This cloud of photons corresponds to Faraday's electromagnetic field, and is responsible for the electromagnetic force between charged particles.

It works like this. A photon that has been emitted by one electron might be absorbed by a second, nearby electron. The photon has no way of recognising its parent electron from any other electron, so this process will inevitably happen to some of the photons in the cloud. The emission of the photon gives a kick to the first electron, and its absorption gives an oppositely directed kick to the second electron. In this way the exchanged photon has transferred some of the energy and momentum of the first electron to the second one, so the trajectory of both electrons will be altered. This is the way that forces operate in quantum theory. They are generated by the interaction of individual particles. It is as though the two electrons are table-tennis players continually batting ping-pong balls back and forth. But because all charged particles exchange photons, the electrons are simultaneously playing photon ping-pong with all the charged particles in their vicinity. The electromagnetic force is a consequence of this exchange of photons between charged particles.

In this way we can visualise Faraday's electric and magnetic fields as they operate at the level of particle interactions. Such pictures are useful stepping stones towards an understanding of physical processes, but what physicists really want are calculations of physical quantities than can be measured in the laboratory. To produce such testable predictions, a new, fully consistent quantum theory of electromagnetism describing electrons, photons and their interactions was required.

Quantum Electrodynamics

Photons, the particles of which light is composed, travel through space at the speed of light, so any useful quantum theory of the electromagnetic force must be consistent with relativity. Dirac took the first step in fusing quantum mechanics and relativity when he devised his relativistic equation of motion for the electron. Building on the success of his equation in the 1930s, he led the way towards a fully quantum theory of electromagnetism. This theory is known as quantum electrodynamics, which is usually abbreviated to QED. But it wasn't until after the Second World War that QED came to fruition. The development of QED into a mature theory was delayed while the leading theoretical physicists were drafted into war work. Many spent the war years working at Los Alamos, the secret laboratory that was established in the New Mexico desert as part of the Manhattan Project – the American programme to develop nuclear weapons.

A theory is worth nothing by itself. Its value derives from using it to perform calculations of quantities that can be compared with experiment. But there were serious technical difficulties that had to be overcome before consistent calculations could be performed with QED. The main problem lay in calculating the effects of the self-interaction of charged particles. The interaction of an electron with its own cloud of photons affects its physical properties, such as its mass and its electric charge. Naive calculations of the size of this effect lead to preposterous results. The same problem exists in classical physics: the electric force obeys an inverse square law, so doubling the distance between two charged particles causes the electric force to fall to a quarter of its original value. Looking at this the other way round, halving the distance between the charged particles quadruples the size of the force. As the distance between the particles decreases further, the size of the force continues to grow, and reducing the distance between the particles all the way to zero will produce a force whose size explodes towards infinity. Similarly, if we consider a charged particle such as an electron to be a spherical ball of negative charge, and try to shrink the size of this sphere down to a point, the repulsive force of one hemisphere for the other hemisphere will increase towards infinity.[17] We cannot produce a sensible theory of an electron if we treat it as point-like, whether the theory is classical or quantum. (This is more of a problem for quantum theory, only because we expect more from quantum physics than we do from classical physics. We expect to be able to understand the fundamental properties of elementary particles in a quantum theory.)

Eventually, three theoretical physicists independently devised methods that would resolve this problem and make QED a viable theory. The trio, two American and one Japanese, were Julian Schwinger, Richard Feynman and Sin-itiro Tomonaga,[18] who all shared the 1965 Nobel Prize in Physics. They worked out how the inconsistencies in the previous treatment of the self-interaction of charged particles could be evaded, and devised techniques that enabled physicists to make reliable calculations. Impossible though it is to calculate the mass and the charge of the electron from first principles in QED, each of them realised that this does not matter. They discovered that as long as they adjusted values for the electron's mass and charge they were using in their calculations to match the electron's mass and charge measured in experiments, they could complete their calculations successfully. Although the theory did not allow the electron's mass and charge to be calculated, these two quantities could be determined in the laboratory. By using just these two pieces of information as inputs into the theory, physicists could in principle calculate the results of any other processes involving electrons.[19] Even so, the calculations still proved to be horrendously complicated.

The Magnetic Field of the Electron

Ampère and Faraday had found that moving electric charges generate magnetic fields. This, of course, applies to electrons. Electrons spin at a constant rate that does not change. This is an intrinsic property of the electron, just as its negative charge is. As the electron spins, its spinning electric charge produces a tiny magnetic field. Physicists call this field the magnetic moment of the electron. Measurements of the magnetic fields carried by electrons in atoms had shown that their spin made twice the expected contribution to the mag-netic field of the atom. One of the great successes of Dirac's equation was that it automatically incorporated the spin of the electron, enabling him to use the equation to calculate the magnetic moment generated by the spin. When he did, the answer came out at exactly 2 (in the appropriate units[20]). In other words, the electron's spin generated twice the magnetic field that would have been produced by an equivalent orbital motion of the electron in an atom – just as had been found in the laboratory.

This calculation was a major triumph. At a stroke, the Dirac equation had solved the mystery and Dirac could explain the experimental result. However, by 1947, many years had passed since the original measurement, and it was now possible to conduct the same experiment with much greater accuracy. Using technology developed during the war, Isador Rabi made new measurements of the magnetic moment of the electron and showed that the magnetic moment was not exactly 2, as predicted by Dirac's equation, but slightly greater. The difference, which became known as the anomalous magnetic moment of the electron, only amounted to about one part in a thousand, but it was definitely real. This discrepancy between theory and experiment highlighted the need for improved calculations based on the embryonic new quantum theory of electromagnetism, QED.

The person who would bring about the reconciliation of theory and experiment was Julian Schwinger, a physicist from New York.[21] Schwinger, born in 1918, was a mathematical prodigy who had written his first paper on quantum physics at the precocious age of sixteen. The direction of his life was determined at an early age by his total absorption in theoretical physics, which became an all-engrossing activity. During the war, when many physicists were occupied with the construction of nuclear weapons, he chose to work instead on the development of radar. The brilliant Schwinger was particularly adept at performing very intricate calculations. Bernard Feld, a colleague of Schwinger's during the war, recalled that if he had a mathematical problem he would go up to Schwinger's office, where Schwinger would be immersed in a calculation. When Schwinger paused, Feld would bring up his own problem:

> I would kind of interrupt him and try to get his attention away from whatever he was doing, and I usually succeeded and not only because I was a pretty persistent guy, but because Julian is a nice guy and if you sort of bother him,

he'll pay some attention to you, and after a while I would get him interested in the particular problem I had in mind, I'd start talking about it and Julian would get interested and then he would go to work on it. He'd get up to the blackboard and I would start taking notes. As Julian worked on the problem, I would be taking notes and sometimes, you know, that could be pretty hectic. I don't know [if you] saw Julian work in those days. Julian is ambidextrous. He has a blackboard technique that uses two hands, and frequently, when he really got carried away, he would be solving two equations, one with each hand, and trying to take notes could be a hectic job. Well, at some point, either we would finish the problem or the dawn would start to break in the eastern horizon, and we would decide it was time to quit and then often, we would go to have breakfast together.[22]

After the war, Schwinger turned his attention to QED. He developed an arsenal of mathematical tools that could be used to take on the formidable calculations of QED. It was Schwinger who was the first to complete a QED calculation of a quantity that could be measured in an experiment. As we have seen, electrons are continually emitting and reabsorbing photons. These processes must be taken into account if we are to make an accurate computation of any of the physical properties of an electron. For instance, if we imagine measuring the electron's magnetic field, then most of the time we will detect the magnetic field of the bare electron, but sometimes we will catch the electron just after it has emitted a photon and before it has had time to reabsorb it. The photon will be carrying some of the electron's energy and momentum, and this will have an effect on the size of the electron's magnetic field.

In 1947 Schwinger calculated the size of this effect. The result of his calculation is usually expressed in terms of what is called the fine structure constant. This is one of the most important constants in the whole of physics: its value determines the strength of the electromagnetic interaction. It gets its name from its earliest appearance in calculations of small electromagnetic effects on the structure of the energy levels of electrons in atoms.[23] The fine structure constant, represented by α, is equal to the square of the electric charge divided by the product of Planck's constant and the speed of light. It combines electromagnetism, quantum mechanics and relativity into one number. And this number is approximately $\frac{1}{137}$. It is a fundamental property of quantum electrodynamics and determines the strength of the electromagnetic force.

Schwinger's virtuoso calculation revealed the size of the biggest QED correction to the predictions of Dirac's equation. He found that there was an additional component to the magnetic moment of an electron, and its size was equal to $\alpha/2\pi$. The resulting prediction was that the magnetic moment of the electron should be slightly bigger than 2: it should have the value

$2[1 + (\alpha/2\pi)]$

which to a good approximation works out as $2 \times 1.001\,162$.

According to Rabi's new measurement, the size of the magnetic moment was somewhere between $2 \times 1.001\,15$ and $2 \times 1.001\,21$,[24] so Schwinger's theoretical value for the magnetic moment precisely fitted the experimentally determined value. This calculation was a landmark event in the history of particle physics. From now on, everyone would have to take QED seriously as a fundamental theory of the electromagnetic force. In the words of the physicist Silvan Schweber:

> The importance of Schwinger's calculation cannot be underestimated. In the course of theoretical developments there sometimes occur important calculations that alter the way the community thinks about particular approaches. Schwinger's calculation is one such instance ... Schwinger had transformed the perception of quantum electrodynamics. He had made it into an effective, coherent and consistent computational scheme.[25]

Schwinger went on to make many more contributions to QED and particle physics in general. He died in 1994 at the age of seventy-six, and is buried at the picturesque Mount Auburn Cemetery in Massachusetts. On his gravestone are inscribed the magic runes: $\alpha/2\pi$.

A Curious Character

Richard Feynman was another New Yorker who was also born in 1918, the same year as Julian Schwinger. These two young independent-minded physicists were very different characters. In the late 1940s the physics community would see them as the two great rivals in the race to understand QED, but whereas Schwinger had a shy, retiring personality, Feynman was a brash extrovert who always said exactly what he thought.

Immediately after completing his PhD dissertation at Princeton University, Feynman left in early 1943 to work as a theoretician on the nuclear weapons programme at Los Alamos, where his brilliance was immediately recognised.

Hans Bethe, the head of the theory division at Los Alamos, recalled that Feynman: 'was very lively from the beginning ... I realized very quickly that he was something phenomenal'.[26]

Feynman was a great solver of puzzles, and when presented with a new one he could not sleep until he had it cracked. At Los Alamos he was also known for his skill with mechanical devices. He had a reputation as a picker of locks and was the person everyone turned to when their mechanical desk calculating machines needed to be repaired. His passion for physics was driven by the same boundless curiosity and his desperate desire to discover the truth behind anything that was supposed to be secret. Feynman has left very entertaining accounts of his activities at Los Alamos in his two volumes of autobiographical tales, *"Surely You're Joking, Mr Feynman!": Adventures of a Curious Character* and *What Do You Care What Other People Think?: Further Adventures of a Curious Character.*

However, though he might sometimes play the joker, when it came to physics Feynman was always deadly serious. After the war he took a position at Cornell University in New York State and resumed his enquiries into electromagnetism and quantum mechanics, the subject of his PhD thesis. Feynman liked to think visually. Many physicists and mathematicians are happy to manipulate algebraic equations and see where they lead, but Feynman was not content unless he could picture what was happening. As he put it himself:

> I dislike all this talk of others [of there] not being a picture possible, but we only need to know how to go about calculating any phenomenon. True we only *need* to calculate. But a picture is certainly a *convenience* and one is not

doing anything wrong in making one up. It may prove to be entirely haywire while the equations are nearly right – yet for a while it helps. The power of mathematics is terrifying – and too many physicists, finding they have the correct equations without understanding them, have been so terrified they give up trying to understand them. I want to go back and try to understand them. What do I mean by understanding? Nothing deep or accurate – just to be able to see some of the qualitative consequences of the equations by some method other than solving them in detail.[27]

This approach led Feynman to his most famous discovery. While he was playing with the equations of QED he developed a pictorial shorthand to help him keep track of all the terms in the horrendously complicated equations. Each diagram was a kind of glyph that represented a mathematical expression, with each part of the diagram corresponding to a mathematical term that formed part of the expression. Feynman devised a recipe that allowed him to translate back and forth between his diagrams and the mathematical expressions that cropped up in QED calculations. This was tremendously useful, because it is much easier to remember the structure of a simple diagram than the exact form of a complicated algebraic expression. But Feynman soon realised that his diagrams were much more than simply memory aids. Not only was each picture equivalent to a chunk of mathematics, it also seemed to depict a physical process – a type of particle interaction.

In an interview in 1966, Feynman recalled how

I was making a lot of these pictures to visualize the various terms and thinking about the various terms [and] a moment occurred – I remember distinctly – when I looked at these, and they looked very funny to me. They were funny-looking pictures. And I did think consciously: 'Wouldn't it be funny if this turns out to be useful, and the *Physical Review* would be all full of these funny-looking pictures? It would be very amusing.'[28]

The first diagram that Feynman published, in 1948, is shown here as Figure 32. It depicts an interaction between two electrons that is mediated by the exchange of a photon between them. One electron emits a photon, which is then absorbed by a second electron. The electrons are shown as solid lines with arrows on them, the photon is shown as a wavy line and the passage of time is upwards. The direction in which the photon was transferred is not distinguished in the diagram (the photon could have been emitted by the second electron and absorbed by the first). Diagrams such as this which do not contain any loops are known as tree diagrams.

Although the diagram is only a schematic representation of an interaction between two particles, it clearly looks very much like the scattering event just described – two electrons approach each other, exchange a photon and then recede

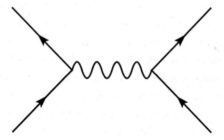

Figure 32 One of the simplest Feynman diagrams, and the first ever published by Richard Feynman. In all such diagrams, time runs upwards.

from each other. But the critical feature of Feynman's pictorial technique is that the diagram snaps together like pieces of a Lego construction kit. Each part of the diagram represents a component of the corresponding mathematical expression. The lines showing the incoming and outgoing electrons can each be written in an equivalent algebraic form. The wavy line representing the exchanged photon corresponds to another algebraic expression. Each vertex where electron and photon lines meet represents a point at which an interaction occurs, and is equivalent to a numerical factor that denotes the strength of the electromagnetic interaction. The number of these factors in the resulting expression is equal to the number of vertices in the diagram. By combining all these pieces, a physicist can take the process depicted in a diagram and reconstruct the full mathematical statement that would result from manipulating the QED equations.[29] When this first Feynman diagram is evaluated, it is found to correspond to a force between the electrons that is an inverse square law, so this diagram is equivalent to the classical physics of Maxwell. This is truly remarkable in itself, but there is much more to the Feynman diagrams.

The expression represented by the diagram in Figure 32 is just one of many that arise in the QED calculation to determine the strength with which a pair of electrons scatter. The full calculation also includes a plethora of other mathematical expressions, each of which can also be represented by a Feynman diagram. Crucially, the same recipe can be used to construct any Feynman diagram: simply identify each algebraic component with the corresponding diagrammatic component and clip them together to form the diagram. Each diagram has the appearance of a physical process, which makes the relationship between the mathematics and the diagrams appear almost miraculous. For instance, one of the diagrams shows the two electrons approaching each other, but before they recede they exchange two photons, as shown in Figure 33. The electrons are again represented as solid lines with arrows on them, and the two photons are represented as the two wavy lines.

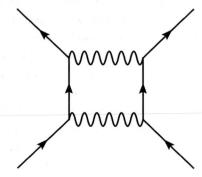

Figure 33 A one-loop Feynman diagram showing the exchange of two photons.

Let's look at another example. Figure 34 shows a photon being exchanged between two electrons. After it is emitted by the first electron, the photon changes into an electron–positron pair, which then mutually annihilate each other and turn back into a photon before the photon is absorbed by the second electron. (The positron is indicated by the leftward-pointing arrow on the curved solid line.)

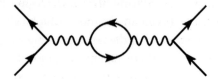

Figure 34 Another one-loop Feynman diagram.

Feynman showed that by thinking in terms of his diagrams, a QED calculation could be performed without
the need for any of the complicated and tedious algebraic
manipulation that had previously been required. All that a
theorist had to do was to construct all the possible diagrams.
Then, using Feynman's recipe, each diagram could be converted into the appropriate mathematical expression. This
process eliminated pages of arduous algebraic hieroglyphics
in which slips would almost certainly creep in, even for the
most meticulous of theoretical physicists. Before Feynman
invented his diagrams, it was only a computational genius
like Julian Schwinger who was capable of completing even
a relatively straightforward QED calculation. Afterwards,
QED became the domain of all trained theoretical physicists. Schwinger ruefully pointed out that Feynman had
given everybody the tools to do QED. In his words, 'Like
the silicon chips of more recent years, the Feynman diagram
was bringing computation to the masses.'[30]

To complete a calculation it was still necessary to evaluate
the mathematical expression corresponding to each diagram,
and for an exact calculation all possible diagrams would have
to be included. However, there is another wonderful feature
of the diagrams: the simpler diagrams give much more significant contributions to the final answer than do the complicated diagrams. They are also much easier to evaluate. It
is possible to get a very accurate answer even if all the complicated diagrams are ignored, because the simplest diagrams
give by far the largest contributions to the final result. This
makes physical sense, because it is much more likely that two
scattering electrons will exchange a single photon, than that
they will exchange two or more photons. The diagram in
which just one photon is exchanged will therefore give the

main part of the result when we calculate the size of the force between the two electrons. The exchange of two photons is next in importance, and so on. In general, the more loops there are in the diagram, the smaller will be its contribution to the final answer. In fact, the diagrams that contain one or more loops represent quantum corrections to the classical result. When they are evaluated they give small deviations from an exact inverse square law. QED thus predicts that at very small distances the force between two electrons is not exactly an inverse square law force, as Maxwell's equations imply. This prediction has been confirmed in the laboratory.

Meditating on Feynman's Mandalas

If Feynman's techniques were to conquer the world of physics, other physicists would first have to understand his methods. Feynman's approach was very intuitive and built on his own, highly individual way of looking at fundamental physics. But although Feynman himself could see how to use the diagrams to navigate a path through the algebraic labyrinth of QED, he had not shown how they related to Schwinger's more orthodox approach. Feynman's flamboyant handwaving style of presentation also didn't help with the initial reception of his revolutionary ideas. During the first explanation of his diagrammatic techniques to an elite group of physicists at a conference in 1947, the elder statesman of quantum physics, Niels Bohr, is reputed to have taken the chalk from Feynman's hand and given him a lecture on the meaning of quantum mechanics.[31]

Freeman Dyson, a young British mathematician, left Cambridge University for America in 1947 to do postdoctoral research at Cornell University. He had decided that the future looked more exciting for theoretical physics than for pure mathematics. Over the course of the next year he got to know Feynman well and was greatly impressed by his ability to perform lightning calculations. Feynman's approach to physics seemed like magic because Feynman 'would write things straight down instead of solving equations.'[32] Dyson was intrigued, and Feynman spent a lot of time with him explaining his methods in detail. But with the instincts of a pure mathematician, Dyson felt uncomfortable. Although Feynman's arguments sounded plausible, it wasn't clear that what he was doing could be justified mathematically.

In August 1948, Feynman, who was travelling to Albuquerque, gave Dyson a lift to a summer school being held at the University of Michigan in Ann Arbor. At the summer school Dyson attended a series of lectures given by Julian Schwinger and had the opportunity to discuss physics with him at length. From Ann Arbor, Dyson took a Greyhound bus to Berkeley in California, and a week later he returned to the East Coast aboard another Greyhound bus. Close contact with both Feynman and Schwinger over the summer paid off. Dyson wrote to his family back in England:

On the third day of the journey a remarkable thing happened; going into a sort of semi-stupor as one does after 48 hours of bus-riding, I began to think very hard about physics, and particularly about the rival radiation theories of Schwinger and Feynman. Gradually my thoughts grew more coherent, and before I knew where I was, I had solved the problem that had been in the back of my mind all this year, which was to prove the equivalence of the two theories. Moreover, since each of the two theories is superior in certain features, the proof of the equivalence furnished incidentally a new form of the Schwinger theory which combines the advantages of both.[33]

Dyson took the insight he had gained on his bus journey and proved that Feynman's graphical methods were mathematically equivalent to Schwinger's rigorous analytical techniques. But whereas Schwinger's techniques were so complicated that they could not be extended to more difficult problems, Feynman's elegant pictorial approach could be developed much further. Dyson published an account of Feynman's methods that would allow others to perform calculations in the same manner, and followed this with a guide to using the Feynman diagrams to calculate physical properties to any desired level of accuracy. This put QED on a firm theoretical basis and enabled it to be tested thoroughly against experiment.

Not of an Age, but for All Time

At the January 1949 meeting of the American Physical Society, Murray Slotnick presented a calculation of the interaction between an electron and a neutron. Robert Oppenheimer, recently the Director of the Manhattan Project, and the most revered figure in American physics, was in the audience. Oppenheimer waited until Slotnick had finished describing his results, then he rose to his feet and imperiously demolished Slotnick's presentation. Oppenheimer dismissed the calculation out of hand stating that it must be in error because it contradicted a theorem that had recently been proved by a mathematician at the Institute of Advanced Study in Princeton. Slotnick was completely taken aback. He knew nothing of this new mathematical result and was unable to respond to Oppenheimer's devastating criticism.

That evening, Feynman arrived at the conference and was told about Oppenheimer's intervention. The mischievous Feynman enjoyed the idea of getting the better of his old boss at Los Alamos, and he suspected that Slotnick might be correct, whatever the pure mathematicians might say. He decided to do the calculations himself to see whether he could corroborate Slotnick's findings. Using his new diagrammatic techniques, Feynman spent a few hours repeating the calculation. In the morning he tracked down Slotnick to compare their results. When Feynman showed him his calculation, Slotnick was stunned. Their results agreed, but whereas it had taken Slotnick two years to complete the painstaking calculation, Feynman had done it in a single evening.[34]

That was the moment that I really knew that I had to publish – that I had gotten ahead of the world ... That was the fire. That was the moment when I got my Nobel Prize, when Slotnick told me that he had been working two years. When I got the real prize it was really nothing, because I already knew I was a success. That was an exciting moment.[35]

Feynman's demonstration that any QED calculation can be represented by a collection of diagrams, combined with Dyson's proof that the procedure was mathematically legitimate, was an immense leap forward. It is hard to do justice to the scale of Feynman's achievement. This is how Dyson saw it:

> Schwinger and Tomonaga had independently succeeded, using more laborious and complicated methods, in calculating the same quantities that Feynman could derive directly from his diagrams. Schwinger and Tomonaga did not rebuild physics. They took physics as they found it, and only introduced new mathematical methods to extract numbers from

the physics. When it became clear that the results of their calculations agreed with Feynman's, I knew that I had been given a unique opportunity to bring the three theories together ... I was careful to treat my three protagonists with equal dignity and respect, but I knew in my heart that Feynman was the greatest of the three and that the main purpose of my paper was to make his ideas accessible to physicists around the world.

Dyson has likened Feynman to Shakespeare, who entranced him with his 'slapdash genius' and who was 'not of an age, but for all time'.[36] His own role he saw as akin to the poet and dramatist Ben Jonson, whose career began as Shakespeare's acolyte, but whose greatest gift to posterity was the publication of the First Folio of Shakespeare's Works in 1623.

Feynman's diagrams have become the talismans of particle physics. Their importance is best exemplified by what is surely the most amazing and involved calculation ever performed. The purpose of the calculation is to set up the most stringent possible test for the theory of QED. The aim is to produce an extremely accurate prediction for the value of a physical quantity which can then be compared with the actual value as measured to high precision in the laboratory. This test will then determine how much confidence physicists should have in the theory and whether it really does embody fundamental truths about the structure of reality.

QED Comes of Age

With Feynman's tools now at their disposal, physicists could improve on Schwinger's calculation of the size of the electron's magnetic moment. The tree diagram in Figure 35

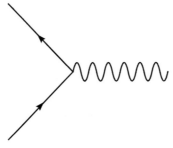

Figure 35 A Feynman diagram representing the largest contribution the magnetic moment of the electron.

represents the most important term in the calculation of the magnetic moment of an electron. The two straight lines with arrows represent the electron. The photon, shown as a wavy line, mediates the interaction between the magnetic field of the electron and the magnetic field of the equipment being used to measure the electron's magnetic field. Evaluating this diagram gives the main contribution to the size of the electron's magnetic moment. This is simply the number 2, in the correct units, which is the result originally derived from Dirac's equation.

Schwinger's calculation that corrected the value of the electron's magnetic moment derived from the Dirac equation is equivalent to evaluating the Feynman diagram in Figure 36. This diagram represents an electron that emits a photon, then interacts with a photon from the magnetic field of the detector, and then reabsorbs the photon it previously emitted. As we have seen, Schwinger's correction to the magnetic moment of the electron was very small: the value derived from the Dirac equation was almost correct. Schwinger calculated that the size of the correction is just $\alpha/2\pi$, where α, as we noted earlier, is approximately

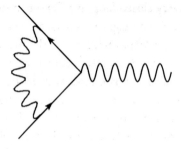

Figure 36 The Feynman diagram that represents the one-loop QED contribution to the magnetic moment of the electron.

equal to ¹/₁₃₇. This term contributes in the region of a mere one-thousandth of the total value of the electron's magnetic moment.

The fine structure constant α gives a fundamental measure of the intrinsic strength of the electromagnetic force. The fact that α is small means that that the electromagnetic force is not too strong, at least for the purposes of performing these calculations. This is ultimately the reason why it is much more probable that a single photon will be exchanged in any interaction than that two or more photons will be

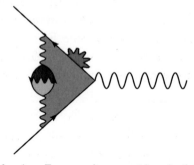

Figure 37 A four-loop Feynman diagram with each of the loops shaded differently.

exchanged. Every closed loop in a Feynman diagram multiplies the value of the diagram by a factor of α. (To illustrate what constitutes a loop, each loop in the diagram shown in Figure 37 is shaded differently.) As we have seen, the diagram that contains a single loop is proportional to α. The next most important Feynman diagrams contain two loops. These diagrams are proportional to α^2. The sum of these diagrams contributes only about one-millionth of the magnetic moment.[37] There are seven such two-loop diagrams, two examples of which are shown in Figure 38.

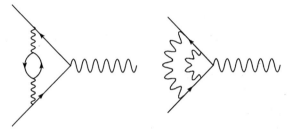

Figure 38 Two examples of two-loop QED Feynman diagrams that contribute to the magnetic moment of the electron.

It had taken Julian Schwinger considerable effort to calculate the first correction to the electron's magnetic moment, and his computation was equivalent to a single Feynman diagram that contained just one loop. The evaluation of seven diagrams, each containing two loops, would be an

Puzzle 7

In terms of the interactions of photons, electrons and positrons, what processes do the two Feynman diagrams in Figure 38 represent?

Answer to Puzzle 7

The diagram on the left in Figure 38 represents a process that is closely related to the one-loop diagram. The electron emits a photon, then interacts with the photon from the magnetic field of the detector and subsequently reabsorbs the photon that it has emitted. However, between being emitted and being reabsorbed the photon dissociates into an electron–positron pair, which subsequently annihilate to re-form the photon.

In the Feynman diagram on the right, the electron emits a photon. It then emits a second photon. The electron then interacts with a photon from the magnetic field of the detector. It then reabsorbs the second photon that it emitted, and finally reabsorbs the first photon that it emitted.

enormous challenge. Dyson was keen to test out Feynman's methods on the calculation. He completed the evaluation of the first of the two-loop diagrams, shown on the left in Figure 38. He then passed on the computation of the other six diagrams to two other young physicists, Norman Kroll and Robert Karplus.[38] Even with the aid of Feynman's techniques, it took the pair two years to complete the evaluation of all seven diagrams.[39]

It was not until 1957 that further experimental advances enabled physicists to test the results calculated by Kroll and Karplus. The accuracy of the new measurements was such that even the tiny corrections obtained from the two-loop Feynman diagrams could be checked against the experimentally determined value. But the outcome was disappointing. There was a discrepancy between the measured value of the magnetic moment and the result calculated by the two theorists. It appeared for a while that QED had failed this latest test. But when new teams of physicists repeated the two-loop calculation, they discovered that Kroll and Karplus had made an error in their evaluation of the Feynman diagrams. As a safeguard against computational slips, the two theorists had performed their calculations independently, but they had met to iron out any differences before publishing their final results.[40] This lack of true independence had allowed critical mistakes to creep through. When the errors in the calculation were corrected, the theoretical QED calculation was found to be in complete agreement with the experimentally determined value. So it turned out that all was well with QED. This episode highlights the formidable challenges involved in the QED calculations.

Engineers test sample materials to destruction in order to determine their strength. Similarly, physicists always attempt to push theories to their limits to find their breaking point. They can then assess the range of validity of their models and see where a better theory is required. They can claim to have a genuine understanding of the way the universe functions only if their theories accurately correspond to the results of measurements made in the laboratory. QED had passed the test of a two-loop calculation, but would it survive the next ordeal? The step to even greater accuracy presented difficulties of another order of magnitude. There are 72 three-loop Feynman diagrams that contribute to the magnetic moment of the electron.[41] And the complications of evaluating each one of these diagrams is far greater than that for any of the previous diagrams. The sum of these diagrams contributes approximately just one billionth of the total value of the magnetic moment. Two examples of the three-loop diagrams are shown in Figure 39.

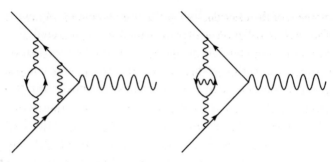

Figure 39 Two of the total of 72 three-loop QED Feynman diagrams that contribute to the magnetic moment of the electron.

Experimental techniques for measuring the electron's magnetic moment were improving all the time. In order to make a more accurate comparison of QED with experiment, the daunting three-loop calculation would have to be made. The calculation was led by the Japanese physicist Toichiro Kinoshita, also based at Cornell University. As a young student in war-torn Tokyo, Kinoshita had been part of Tomonaga's team that had helped to elucidate QED in the 1940s. With better evaluation techniques and the assistance of the first generation of electronic computers, it took teams of theoreticians over ten years to complete the three-loop calculation. Amazingly, comparison of the three-loop calculation with the experimentally determined value showed complete agreement. The QED prediction fitted the measurement to nine places of decimals. This level of accuracy – one part in a billion – is equivalent to one second in about thirty-two years.

Of course, Kinoshita and his colleagues were not tempted to rest on their laurels after the success of their three-loop evaluation of the electron's magnetic moment. There are 891 four-loop Feynman diagrams that contribute to the magnetic moment of the electron; two of them are shown in Figure 40. The total contribution of all these diagrams is a minuscule one-trillionth of the total magnetic moment. But if a discrepancy between the QED prediction and the experimental measurement appears, then it would indicate the limits of QED and point physicists in the direction of an even better theory. Alternatively, if they continue to tally at this level of precision it will give physicists even greater confidence that QED is an incredibly accurate representation of reality. With improved mathematical techniques and the aid of modern computers, all 891 four-loop diagrams have now been evaluated.

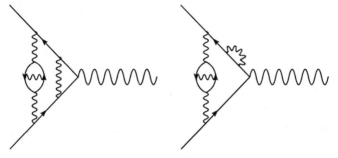

Figure 40 Two examples from a total of 891 four-loop QED Feynman diagrams that contribute to the magnetic moment of the electron.

Experimental procedures for measuring the electron's magnetic moment have advanced in step with the progress in the theoretical calculation. Currently the most precise measurement comes from studying the properties of a single electron confined in a magnetic field, an arrangement known as a Penning trap.[42] Initially the trap contains many electrons, but one by one the electrons are expelled until just a single electron remains. This electron orbits around in the magnetic field in an arrangement that is similar to a giant hydrogen atom, but here the electron is confined by an external magnetic field rather than by its interaction with the atomic nucleus. The strength of the external fields can be controlled very precisely, and their effect on the energy levels available to the electron can be calculated. The electron is suspended in this way for several months at a temperature that is just a fraction of a degree above absolute zero. By altering an electric field perpendicular to the plane of the electron's orbit, the size of the gaps between the energy levels in the trap can be determined with an incredible accuracy. Quantum mechanical calculations show that these energy gaps are proportional to the QED corrections to the electron's magnetic moment. Measuring the energy gaps gives a direct way to determine the anomalous magnetic moment of the electron to an extraordinary precision.[43]

The most recent experimental result was published by a group at Harvard University in 2006. Their measurement gives the following value for the electron's magnetic moment:

$$2 \times 1.001\,159\,652\,181$$

with an estimated uncertainty of less than one part in a trillion.[44] This is an incredibly precise measurement of a physical quantity. It is equivalent to the measurement of a distance equal to the Earth's circumference to an accuracy to within the width of a very fine human hair.

So how does the theoretical prediction stand up against this most recent measurement from the laboratory? The physics community has now been undertaking the QED calculation for around sixty years and the latest theoretical value for the magnetic moment of the electron includes contributions from about a thousand Feynman diagrams, that is,

---

Puzzle 8

See if you can draw a couple of the five-loop Feynman diagrams. Just to make clear what constitutes a loop, each of the loops in the four-loop diagram shown in Figure 37 is shaded differently.

---

all the diagrams with up to four loops.[45] To give a numerical statement of the value of the magnetic moment predicted by the theory, we need a precise value for α, the fine structure constant, and this value must be determined by an independent experiment. Fortunately it is possible to determine α to within a few parts in a trillion from the spectroscopy of caesium and rubidium atoms. If we do this and combine the result with the QED calculation, the resulting theoretical value for the electron's magnetic moment is found to be

$$2 \times 1.001\,159\,652\,178$$

with an accuracy estimated to be about nine parts in one trillion. This accuracy is limited by the precision of the spectroscopic measurements that have been used to determine α, and not by any inaccuracies in the QED calculation itself. This level of precision implies that the theoretical prediction for the size of the electron's magnetic moment lies somewhere between

$$2 \times 1.001\,159\,652\,169 \quad \text{and} \quad 2 \times 1.001\,159\,652\,187$$

If we compare these numbers with the latest measured value for the electron's magnetic moment, given above, we can see that the measured value is in the middle of the range given by the theoretical calculation. This shows an incredible agreement between theory and experiment. To an accuracy of within ten parts in one trillion, the prediction of QED has been confirmed. And that is an accuracy equivalent to one second in around 3,200 years. The increasing levels of accuracy achieved by taking into account successive sets of Feynman diagrams is shown in Figure 41.

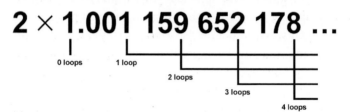

Figure 41 The levels of accuracy at which each set of Feynman diagrams begin to contribute to the calculation of the magnetic moment of the electron.

There is no indication from this or any other experiment that QED has been stretched beyond its range of applicability. This is certainly one of the great triumphs of modern science. QED must be telling us something significant about the fundamental structure of reality for its predictions to have been so precisely confirmed. The agreement between theory and experiment goes way beyond what the pioneers of QED, such as Schwinger and Feynman, had anticipated. They had expected that QED would soon be replaced by a better theory.

So where do we go from here? Experimental techniques are improving all the time. An even more precise measurement of the electron's magnetic moment will soon be available. On the theoretical side, the enormous task of evaluating all the Feynman diagrams with five loops has already begun. It took the great Julian Schwinger months to evaluate the single expression corresponding to the first and simplest diagram with just one loop. The octogenarian theoretician Kinoshita and his younger colleagues around the world are bracing themselves for the awesome task of evaluating all 12,672 Feynman diagrams with five loops, each of which is more complicated than any of the diagrams with four loops. In addition, at this level of accuracy it will also be necessary to include contributions from Feynman diagrams containing loops of particles other than electrons. These particles will include the quarks, a family of particles whose story will be told in a later chapter.

In this chapter we have looked at the work of a number of great theoretical physicists. Each has had his own distinctive approach to physics, and each has played a vital role in the development of QED. But their contributions to science have turned out to be even more significant than this. Much of modern theoretical physics is built on their ideas. QED is not only a phenomenally successful theory of the electromagnetic interaction, it has proved to be an exceptionally

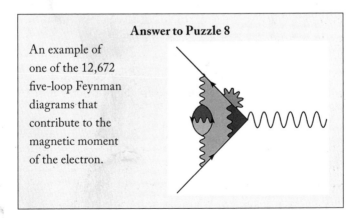

Answer to Puzzle 8

An example of one of the 12,672 five-loop Feynman diagrams that contribute to the magnetic moment of the electron.

good model for the construction of theories that describe the other two forces that play a role in particle physics. These two forces are the strong force and the weak force. Both are critical to the structure of our universe. We will be taking a look at the modern theory of the strong force later, but first we must journey to the stars to see the weak force in action.

Chapter Five

WE ARE STARDUST!

We are stardust, we are golden,
And we've got to get ourselves back to the garden.
We are stardust, billion year old carbon,
We are golden.

Joni Mitchell, 'Woodstock' (1969)

The Age of the Earth

The great Victorian physicist William Thomson, who was later raised to the peerage as Lord Kelvin, claimed that the Earth could be no more than one hundred million years old. He arrived at this figure by assuming that the Earth had formed as a molten ball and working out how long it would take to cool from this state to its present temperature. One hundred million years was certainly a very long time in comparison to the traditional Christian belief that the universe had been created just a few thousand years earlier. But it still presented a serious problem for scientists because it didn't appear to leave sufficient time for the glacial pace of geological processes or for the evolution of life as described in Darwin's theory of natural selection. At around the same time, the German physicist Hermann von Helmholtz tackled the related problem of the age of the Sun and derived a figure that was comparable to Kelvin's. Helmholtz calculated the length of time it would take the Sun to contract under its own gravitation from a cloud of gas into a spherical ball of its observed size, assuming the Sun's energy was being generated by this gravitational collapse.[1]

Both Kelvin's and Helmholtz's calculations were built on very shaky foundations, as Darwin's great champion T.H. Huxley pointed out at the time. These physicists had based their calculations on energy sources whose existence was known in the Victorian era, such as fossil fuels. By the end of the century, a new source of energy had been discovered that would lead to a revolutionary transformation of physics in the twentieth century. In 1896, Henri Becquerel noticed that some photographic plates had become fogged when they were left in a desk drawer alongside a lump of a mineral that contained the element uranium. After investigating this phenomenon he deduced that the uranium was emitting a new type of ray that was affecting the photographic plates. Becquerel had stumbled across radioactivity. This completely unexpected discovery would launch the era of nuclear physics.

The physics of the nucleus and the energy generated in nuclear processes would eventually explain why Kelvin and

Helmholtz got their calculations so badly wrong. It is the continuous release of energy from radioactive elements such as uranium within the Earth that has kept the Earth's innards molten for billions of years and prevented it from freezing over like a gigantic snowball. Remarkably, radioactivity itself has also provided us with a precise determination of the age of the Earth. Analysis of the accumulated decay products of uranium, in rocks from the Earth, the Moon and from meteorites, has yielded a well-established age for the Earth and the rest of the solar system of about 4.54 billion years. As we shall soon see, nuclear physics also enables us to understand the true source of the Sun's energy and explains how it is still shining steadily after four and a half billion years.

The World Set Free?

The professor went on after a little pause. 'Why is the change gradual?' he asked. 'Why does only a minute fraction of the radium disintegrate in any particular second? Why does it dole itself out so slowly and so exactly? Why does not all the uranium change to radium and all the radium change to the next lowest thing at once? Why this decay by driblets; why not decay *en masse*? … Suppose presently we find it possible to quicken that decay?'

The chuckle-headed lad nodded rapidly. The wonderful inevitable idea was coming. He drew his knee up towards his chin and swayed in his seat with excitement. 'Why not?' he echoed, 'why not?'

The professor lifted his forefinger.

'Given that knowledge,' he said, 'mark what we should be able to do! We should not only be able to use this

uranium and thorium; not only should we have a source
of power so potent that a man might carry in his hand the
energy to light a city for a year, fight a fleet of battleships
or drive one of our giant liners across the Atlantic; but we
should also have a clue that would enable us to quicken the
process of disintegration in all the other elements, where
decay is still so slow as to escape our finest measurements.'

H.G. Wells, *The World Set Free* (1914)[2]

The earliest discoveries of radioactivity suggested that the
atom contained a source of energy that far outstripped anything previously experienced. This energy is released gradually, but over a period of time a radioactive element such as
radium will emit vastly more energy than could be obtained
by burning a lump of coal of the same size. At least to a few
far-sighted individuals, the implications were clear. Inspired
by the latest advances in physics, the great science fiction
writer H.G. Wells published an account of a nuclear war and
its aftermath called *The World Set Free*. Incredibly, this was
on the eve of the First World War, over thirty years before
nuclear weapons became a horrific reality. But the origin
of the vast outpouring of energy witnessed in radioactivity
would become clear only when the structure of the nucleus
was understood. And it would be Rutherford and his team
who would reveal this structure.

James Chadwick was born in 1891 in the Cheshire village of
Bollington, just south of Manchester. His father was a poverty-
stricken cotton-spinner who moved the family to Manchester
in the hope of improving their fortune, but without success.
Despite these difficulties, James worked hard at school, and
with the help of a scholarship he won a place at Manchester
University. By 1911 he had become a member of Rutherford's

team of researchers working on various aspects of radioactivity. In 1912, Hans Geiger, who had performed the critical experiment with Ernest Marsden that revealed the nucleus, left Rutherford's team to lead a group of researchers in Berlin. The following year Chadwick joined Geiger in the German capital. This was a golden period for physics in Berlin, where both Einstein and Planck held professorships at the university. However, it proved to be a bad move for Chadwick. When war broke out a year later he was interned as an enemy alien in a prisoner-of-war camp near Berlin. He remained in the camp for most of the war, and although he managed to set up a laboratory in stables where he could perform some chemistry experiments, conditions were dreadful, and he later recalled that it had nearly killed him.[3]

In 1919, Rutherford left Manchester and returned to Cambridge as director of the Cavendish laboratory. Later the same year, Chadwick was repatriated and rejoined Rutherford in Cambridge. During the 1920s, Rutherford's team considered various possible ways in which the existence of the neutron – a neutral sister particle to the proton – might be established, and Chadwick undertook several experimental searches, but with no success.

In 1930, two German researchers, Walter Bothe and Herbert Becker, discovered that they could generate highly penetrating radiation by firing alpha particles at the light metal beryllium. They assumed that this must be gamma radiation, which is electromagnetic radiation similar to visible light, but at the very-high-energy end of the spectrum. At the beginning of 1932, news arrived in Cambridge that when this radiation was aimed at a block of paraffin wax, protons were ejected from the paraffin. The molecules in paraffin contain a lot of hydrogen atoms, and this was the source of the protons.

The intense radiation was blasting the nuclei of the hydrogen atoms, which consist of single protons, right out of the paraffin. It was clear to Rutherford and his colleagues that the identification of the radiation as gamma rays must be invalid. They did not believe that gamma radiation could be remotely powerful enough to knock protons out of the wax; furthermore, they had a much better candidate for the radiation.

A decade of thinking about Rutherford's hypothetical neutral particle had prepared their minds: they were on the look-out for the neutron. The vital clue was their realisation that the neutron would have a similar mass to the proton. While it was unrealistic to expect a gamma-ray photon to knock a proton out of paraffin, it was perfectly reasonable to expect that a proton could be knocked out by a second particle with a similar mass. (Imagine playing snooker or pool with a marble as the cue ball instead of the usual white ball.) Chadwick seized his opportunity. In a month of feverish activity, he performed a series of experiments that tied down the properties of the radiation and provided the definitive proof that the neutron had been found. In February 1932 Chadwick published his paper announcing the discovery of the neutron. He was rewarded with the 1935 Nobel Prize in Physics.

The neutron was the key to the structure of the nucleus, and its discovery transformed our understanding of the composition of matter. In just over a decade, hastened by the urgencies of the Second World War, the awesome power of the nucleus would be released. It was now apparent that the nucleus of an atom is composed of a tightly bound collection of two types of particle: protons and neutrons. In a relatively light nucleus such as that of a carbon atom, there are approximately equal numbers of protons and neutrons. Carbon is element number 6, and there are three main forms of the carbon nucleus: carbon-12, carbon-13 and carbon-14, which are known as different isotopes of carbon. They are represented symbolically as ^{12}C, ^{13}C and ^{14}C. Each of the three types of carbon nucleus contains six protons, so there will be six electrons in any neutral carbon atom. But each type of nucleus contains a different number of neutrons. For instance, ^{12}C, the nucleus from which 99 per cent of carbon is formed, contains six neutrons, so its nucleus is composed of a total of twelve nucleons. (Nucleon is the collective name for neutrons and protons.) This is the reason for the superscript 12 in its symbol. A ^{13}C nucleus contains seven neutrons, giving a total of thirteen nucleons in the nucleus. In heavy atoms such as uranium, which is element number 92, there are always more neutrons than protons in the nucleus. For instance, a nucleus of uranium-238 (^{238}U) contains 92 protons and 146 neutrons.

Taking into account the existence of the neutron, it is now easy to interpret the nuclear reaction observed by Bothe and Becker. An alpha particle is a helium nucleus, composed of two protons and two neutrons. A beryllium nucleus is composed of four protons and five neutrons. When Bothe and Becker fired alpha particles at their beryllium target, an alpha particle could be captured by a beryllium nucleus, which would then spit out a neutron to leave a stable nucleus of carbon, containing six protons and six neutrons. This reaction can be represented as follows

$$^{4}\text{He} \quad + \quad ^{9}\text{Be} \quad \rightarrow \quad ^{12}\text{C} \quad + \quad n$$
$$(2p, 2n) \quad (4p, 5n) \quad (6p, 6n) \quad (n)$$

where p represents a proton and n represents a neutron. The total number of protons and the total number of neutrons remains the same after the reaction as it was before the reaction.

The various nuclear processes involved in radioactivity can also be understood at the level of particle interactions. A nucleus will undergo radioactive decay if it can fall into a lower energy state. The simplest form of radioactive decay is gamma decay, in which a nucleus loses energy by emitting a gamma-ray photon. This is the nuclear equivalent of an electron in an atom falling down to a lower energy level by emitting a photon, but because nuclear processes involve much higher energies, the photon released by a radioactive nucleus is a very-high-energy gamma-ray photon rather than a much lower-energy photon of visible light.

The particles that form the nucleus also provide an explanation for alpha and beta decay. Alpha decay tends to occur in heavy nuclei. It occurs, for instance, in the unstable uranium-238 nucleus. This nucleus can discard some of its energy and become a more stable nucleus by emitting an alpha particle, consisting of two protons and two neutrons bound together. What remains is a nucleus containing 90 protons and 144 neutrons. Alpha decay converts uranium-238 into a nucleus of element number 90, which is called thorium.

Beta decay is rather different. Of the different carbon isotopes, the nuclei of both carbon-12 and carbon-13 are stable, but carbon-14 is radioactive – it undergoes beta decay. The radioactivity of carbon-14 is utilised in carbon dating, a valuable tool that enables archaeologists to calculate the ages of materials that are up to a few thousand years old. When it decays, a carbon-14 nucleus, which contains six protons and eight neutrons, emits a beta particle, which is identical to the electron. The nucleus that remains afterwards contains seven protons and seven neutrons and is therefore an isotope of element number 7, nitrogen. Along with the emission of the electron, one of the neutrons has been magically transformed into a proton. Notice that electric charge is conserved in the process of beta decay – as it must be in all interactions. The electric charge of the neutron is zero. The electric charge of the proton is +1 and the electric charge of the electron is −1, so the total electric charge remains zero:

$$n \rightarrow p^+ + e^-$$

(electric charge)   0   +1   −1

It would be wrong to imagine that the neutron actually contains a proton and an electron. In a related type of radioactive decay, a proton can emit a positron and turn into a neutron. Again, the proton does not contain a positron and a neutron; rather, the proton and the neutron are being transformed into each other in these processes.

In the age of the particle accelerator it has become customary to quote the energy of a particle in electronvolts (eV). Because mass is just another form of energy, the mass of a particle is also usually stated in electronvolts. This is convenient when we are considering the production of particles in accelerators. One electronvolt is the energy that an electron, or any other particle carrying one unit of electric charge, gains when it is accelerated through a one-volt electrical circuit. Chemical reactions typically involve energies of a few electronvolts. This is why batteries have a voltage in this range, because the process by which they produce an electric current is a chemical one. Further up the scale, 1,000 electronvolts is written as 1 keV, where keV stands for kiloelectronvolts. The energy released in a nuclear process typically involves about a million times as much energy as a chemical reaction, so it is generally quoted in MeV (megaelectronvolts). 1 MeV equals 1,000,000 electronvolts. When a uranium-238 nucleus undergoes alpha decay, the alpha particle that is released is always emitted with the same sharply defined energy: 4.2 MeV. This is about a million times as much energy as is involved when an electron undergoes a transition between atomic energy levels. One billion electronvolts (i.e. one thousand million electronvolts) is written as 1 GeV (gigaelectronvolt). 1 GeV is slightly more than the mass of a proton. One trillion electronvolts, or 1,000 GeV, is written as 1 TeV, which stands for teraelectronvolts. The Large Hadron Collider is currently smashing protons together with a combined energy of 8 TeV, and eventually it will reach an energy of 14 TeV – over seven times the record previously achieved in a particle accelerator.

The Elusive Neutrino

In 1914, just a few years after Rutherford had first identified the atomic nucleus, a feature of the beta decay process was discovered that presented physicists with a serious puzzle. Unlike alpha particles, which were always emitted with a sharply defined energy, beta particles were released with a continuous range of energies, up to a maximum energy that was a characteristic of each particular nucleus. This was hard to understand, because if the nucleus was undergoing a transition from a higher energy level to a lower one, the amount of energy released should always be the same, and the decay products should always carry away the same total amount of energy. Conservation of energy requires that the difference in energy between the original and final energy levels of the nucleus must equal the total amount of energy released.[4] One possibility was that the energy released was being shared between the electron and a second particle that was escaping without being detected – a gamma-ray photon, for instance. If that were so, then as long as the total energy of the electron and the photon was equal to the difference in the two energy levels, energy would still be conserved.

This plausible explanation could be checked by measuring the amount of heat produced by the beta decay of a radioactive isotope, which would determine whether the missing energy was being released as gamma-ray photons. Even if every individual gamma-ray photons could not be detected, the total amount of energy they were carrying could be measured. In 1927, just such an experiment was performed using a radioactive isotope of bismuth.[5] Researchers placed the bismuth sample in a well-insulated container so that they could monitor its temperature. They found that the temperature did indeed rise, as the energy released by the gradual beta decay of the bismuth heated the sample. When the total number of beta decays occurring during the period of the experiment was calculated, it became clear that the temperature rise could be completely accounted for by the energy that was released with the electrons emitted by the decaying nuclei. There was no additional rise in temperature that could be attributed to the absorption of energy from any other undetected particles that might also be released in beta decay. It was now certain that no gamma rays, or other known particles, were being emitted along with the electrons. It seemed as though the electrons must be taking away all the energy produced in beta decay, but that this energy amounted on average to only about half what was expected from the difference in the two nuclear energy levels.[6] The mystery remained.

At this time, physics was in the middle of the turmoil produced by the emergence of quantum mechanics. Physicists were in the process of completely restructuring their subject, and it seemed that many long-cherished principles might have to be abandoned. In this revolutionary atmosphere, it appeared to Niels Bohr and other leading physicists that the beta-decay results were indicating that the conservation of energy might be another principle that should be jettisoned. However, in 1930 Wolfgang Pauli offered a different explanation. He suggested that the electron was indeed being emitted along with another particle, but that this was not a gamma-ray photon or any other known particle.[7] Pauli proposed that it must be a new fundamental particle. This was a radical proposal in 1930, as the only fundamental particles whose existence had been established were the electron, the proton and the photon. The fact that Bohr preferred to entertain the possibility of a violation of the law of energy conservation shows just how radical the prediction of a new particle was considered to be.

The Italian physicist Enrico Fermi was one of the few twentieth-century physicists to do really important work both as a theorist and as an experimenter. On 2 December 1942 he constructed the world's first nuclear reactor in a squash court under the American football field at the University of Chicago. He initiated the age of nuclear power by setting up a nuclear fission chain reaction and allowing it to continue for about half an hour before shutting it down. Ten years before this demonstration of the possibility of power generation from nuclear processes, Fermi had made important theoretical advances in the understanding of beta decay. He recognised that beta decay could not be explained by either the electromagnetic or the strong nuclear force, so it must be the result of a new force of nature, a force that was much weaker than the electromagnetic force. This is the force we now know as the weak nuclear force, or simply the weak force. Fermi also invented a name for Pauli's new particle. He called it a neutrino – the 'little neutral one'.[8]

Pauli identified a number of characteristics the neutrino must have. It was clear that neutrinos must have a very small mass, and most obviously that their interactions with matter must be very weak – which was why they had evaded detection, and why they were able to escape and carry off some of the energy released in beta decay. As we have seen, the conservation of electric charge in beta decay could be accounted for without the need to invoke the neutrino, so the neutrino must have zero electric charge – which is why it does not interact via the electromagnetic force. It was also clear that neutrinos do not interact with atomic nuclei, otherwise they would not be able to pass through matter so easily without undergoing any interactions. This means that, like electrons, they do not feel the strong force.

We can now include neutrinos in the formula for the beta decay process:

$$n \rightarrow p^+ + e^- + \nu$$

(electric charge) 0 +1 −1 0

where the neutrino is represented by the Greek letter nu (ν).

Neutrinos interact with other particles only via the weak force, and these interactions really are incredibly weak. A quick comparison with alpha and beta radiation will give us some idea of just how weak these interactions are. Alpha particles – helium nuclei – interact by both the electromagnetic and the strong force. An alpha particle emitted in the radioactive decay of an unstable nucleus, with an energy of a few million electronvolts, will be stopped by a single sheet of paper. An electron with a similar energy emitted in beta decay will be stopped by a lead plate one millimetre thick, and a gamma-ray photon will typically penetrate about one centimetre of lead. A typical neutrino,[9] however, could be expected to pass through a solid block of metal such as gold or lead with a depth of several light years before interacting. In other words, it could travel all the way through solid gold to the nearest stars, as though the gold were transparent.

The nature of these elusive beasts was elegantly summarised by John Updike in his poem 'Cosmic Gall', which begins:

> Neutrinos, they are very small.
> They have no charge and have no mass
> And do not interact at all.
> The earth is just a silly ball
> To them, through which they simply pass,
> Like dustmaids down a drafty hall
> Or photons through a sheet of glass.[10]

So how do we know that these elusive particles really exist? Physicists were never going to fully accept a hypothetical particle that could not be seen in the laboratory, and detecting a neutrino was certainly not going to be easy. In 1951 Fred Reines and Clyde Cowan decided to take up the challenge. In Reines' words, 'Why not, I thought, pick an important but "impossible" problem and solve it – that would really be something!'[11] Spotting a single individual neutrino would, indeed, be impossible. But if neutrinos could be produced in sufficient quantities, a few very rare interactions might be detected, and Reines and Cowan realised that a new source of copious neutrinos had recently become available – the nuclear reactor.[12]

The nuclei produced by the fission of uranium nuclei in a nuclear reactor are highly radioactive, and many of them undergo beta decay, so reactors are continuously emitting an enormous flux of neutrinos. Reines and Cowan estimated that every second around ten trillion (10^{13}) neutrinos would escape from every square centimetre – an area the size of a postage stamp – of the surface surrounding the core of a nuclear reactor.[13]

In 1956, Reines and Cowan built targets containing cadmium chloride solution that were designed to intercept neutrinos escaping from the Savannah River reactor in South Carolina. For the first time they successfully spotted small numbers of neutrinos, which were detected as they interacted with protons in the target solution,[14] in a process known as inverse beta decay:

$$\nu + p \rightarrow n + e^+$$

Here a neutrino (ν) interacts with a proton (p); the proton is converted into a neutron (n), and the neutrino is simultaneously converted into a positron – the antiparticle of the electron, represented here by the symbol e^+. Almost immediately, the positron that is produced meets an electron in the target solution, and they annihilate with the emission of a pair of gamma rays. These gamma rays are easily spotted in a detector because they each have an energy equal to the mass of the electron (and the positron), of 511 keV. The reason for using cadmium chloride as the target was that the neutron would then be absorbed by a cadmium nucleus in the target solution to form a radioactive isotope of cadmium that decays within a few microseconds, with the emission of a gamma ray with an energy of around 8 MeV. The detection of two gamma rays, each with an energy close to 511 keV, followed immediately by the detection of an 8 MeV gamma ray, was the characteristic signal that Reines and Cowan were looking for and would allow them to deduce that they were indeed witnessing the interaction of the neutrino with a proton in their target.

With their apparatus they were able to detect about three neutrinos per hour. In a series of elegant experiments they demonstrated definitively that the particles they were detecting were neutrinos, and not some other type of particle. Among other things, they showed that their results were not affected by encasing their apparatus within a lead shield. If the activity had been produced by something else, such as gamma rays or neutrons, interacting with the protons, the activity would have decreased markedly when the lead shielding was added, as gamma rays and neutrons would have been absorbed by the lead before reaching the detector solution. The only reasonable conclusion was that they were, indeed, genuinely observing the incredibly rare interactions of neutrinos.[15] Almost forty years later, in 1995, Fred Reines was rewarded for his work with a Nobel Prize in Physics.

Since the groundbreaking work of Reines and Cowan, there have been major advances in neutrino detection. There are now a number of experimental facilities around the world that are specifically designed to spot the occasional interaction of a neutrino. Many of these detectors are built deep underground in mines to reduce the amount of background radiation from cosmic rays. One of these is the Kamiokande neutrino observatory, built in 1983 in the Mozumi mine in Japan, a kilometre beneath the Earth's surface. The original detector consisted of 4,500 tonnes of ultra-pure water. By 1995 the facility had been upgraded to the Super-Kamiokande with tanks containing 50,000 tonnes of ultra-pure water. Recent observations at Super-Kamiokande have revealed that neutrinos have a tiny mass, far, far less than the mass of an electron. The mass of an electron is about 511 keV, whereas the mass of a neutrino is just 2 eV. This is minuscule, less than the energy gained by an electron passing through a circuit connected to a two-volt battery. The electron, which is usually considered to be a lightweight particle, is at least a quarter of a million times as heavy as a neutrino, and the proton is at least half a billion times as heavy as a neutrino.

Fusion

The weak force may be weak, but it would certainly be wrong to assume that this means it is unimportant. When an electron interacts with a proton via the electromagnetic force, the electron remains an electron and the proton remains a proton. Energy and momentum are exchanged between the two particles as they interact, but the two particles keep their identities. The weak force, on the other hand, is like the philosopher's stone of the alchemists: it can change the identity of matter. It can transform base matter into gold. Without the weak force there would be no periodic table, as all matter would consist of the simplest atom, hydrogen. Without the weak force the universe would be devoid of all complex and interesting phenomena. There would be no life.

Look up at the awesome spectacle of a dark night sky. The bejewelled heavens are filled by a thousand points of light, each one an immense, roiling nuclear cauldron blasting intense radiation deep into interstellar space. The nineteenth-century philosopher Auguste Comte insisted that the chemistry of the stars would always be hidden from us, that it was impossible to gain such knowledge. But as we have seen, despite Comte's scepticism it wasn't too long before this puzzle was solved and the spectral signatures of the elements were identified in starlight. A century later, physicists were able to take the story much further by determining exactly how the chemical components of the stars came to be there in the first place.

There are about twenty-five elements that are essential for life. Our DNA is formed from atoms of carbon, nitrogen, oxygen, phosphorus and hydrogen. To form proteins, we must add atoms of sulphur to this list. We need iron to form the haemoglobin that transports oxygen around our body. Calcium is required for our bones. Sodium and potassium are needed for the transmission of impulses along our nerves. Atoms of iodine are found in the hormone thyroxin. Plants utilise magnesium in chlorophyll molecules that enable their leaves to capture energy from sunlight. It is an amazing fact that, with the exception of the simplest atom, hydrogen, all the atoms in our bodies were forged in the alchemical laboratories that we know as stars. We truly are stardust, as Joni Mitchell sang so magically.

Working out how the chemical elements are synthesised in stars was one of the great scientific achievements of the twentieth century. In 1920 the British astrophysicist Sir Arthur Eddington was the first to realise that the fusion of hydrogen nuclei to form helium nuclei might be the source of the energy that keeps the Sun and other stars shining.[16] The nucleus of a hydrogen atom is a single proton, whereas the nucleus of a helium atom consists of two protons and two neutrons. Eddington, an enthusiastic supporter of Einstein's theories of relativity, realised that if a helium nucleus could somehow be forged from four protons, then the difference between the mass of four protons and the mass of the helium nucleus would be liberated as energy according to Einstein's famous formula $E = mc^2$. The mass of a proton is 938.3 MeV. Four times this mass is 4×938.3 MeV = 3,753 MeV, to the nearest whole number. The mass of a helium nucleus is just 3,727 MeV. The difference in mass between the four protons and the helium nucleus is therefore about 26 MeV. So if four protons could indeed be converted into a helium nucleus, then about 0.7 per cent of the total mass of the protons would be converted into energy.

But how can four protons form a helium nucleus when the helium nucleus consists of two protons and two neutrons? The answer was worked out by Hans Bethe and Charles Critchfield in 1938. Two protons very close together will feel an attraction due to the strong force. However, protons carry a positive electric charge, so there is a big electro-magnetic repulsion between two protons. This makes it very difficult for two protons to get close enough for the strong force to play a role in their interaction. The two protons can approach each other sufficiently closely for a fusion reaction to occur only if they are moving at extremely high veloci-ties. This corresponds to a very high temperature, as exists at the centre of a star like the Sun, which is about 15 mil-lion degrees Kelvin. Crucially, although the strong force will bind collections of protons and neutrons and will hold a single proton and a single neutron together, or two protons and a neutron, it is not quite strong enough to form a nucleus from just two protons. Without the additional 'glue' provided by a neutron, the electromagnetic repulsion between two protons is a little too strong. This is of great significance for the fusion reactions within a star. It means that for the first step along the road to a helium nucleus, two protons must be forced together in a high-speed collision, and at the exact moment of impact one of the protons must simultaneously undergo the weak interaction known as inverse beta decay.[17] The proton is converted into a neutron, with the emission of a positron and a neutrino. The other proton and the newly formed neutron then bind together to form a nucleus of heavy hydrogen, an isotope known as deuterium. We can represent this nuclear reaction as follows:

$$p + p \rightarrow (p, n) + e^+ + \nu$$

or, completely equivalently,

$$^1H + {}^1H \rightarrow {}^2H + e^+ + \nu$$

Here ^1H represents a proton, the nucleus of hydrogen, ^2H represents a nucleus of deuterium or heavy hydrogen, e^+ represents a positron and ν represents a neutrino.

This is the critical step in the synthesis of the elements and the liberation of energy in stars. Without the weak interaction that converts one of the protons into a neutron, there would be no matter in the universe other than protons and electrons – the universe would consist of hydrogen and nothing else. However, the extreme weakness of the weak force makes this step exceedingly slow. Just one collision in ten billion trillion (10^{22}) between two protons in a star such as the Sun results in a fusion reaction.[18] Typically, under the conditions that prevail in the interior in the Sun, a proton will race around, ricocheting off other protons, for about ten billion years before it undergoes this reaction, so this is the bottleneck that determines the overall rate of the conversion of hydrogen into helium. Although any individual proton will typically wait almost an eternity before undergoing this reaction, the quantity of protons in a star is vast and the production of deuterium proceeds steadily throughout the star's lifespan.

The next step in the fusion process happens almost immediately. Within one second, the deuterium nucleus captures another proton to form a nucleus of light helium containing two protons and one neutron. This is a nucleus of helium-3. We can represent this reaction as follows:

$$(p, n) + p \rightarrow (p, p, n) + \gamma$$

or equivalently

$$^2H + {}^1H \rightarrow {}^3He + \gamma$$

Here, gamma γ represents the gamma-ray photon that carries away most of the energy produced in the reaction; γ is the standard symbol for a photon.

On average it will take another million years for the helium-3 nucleus to meet another helium-3 nucleus in the star and undergo the following nuclear reaction that results in the production of helium-4:

$$(p, p, n) + (p, p, n) \rightarrow (p, p, n, n) + p + p$$

or equivalently

$$^3He + {}^3He \rightarrow {}^4He + p + p$$

with two protons being released back into the plasma soup. Overall, the result of these reactions is the conversion of protons into helium-4 nuclei (see Figure 42), just as Eddington had originally proposed.

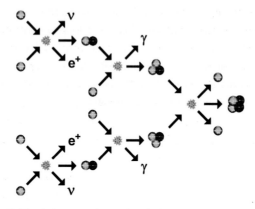

Figure 42 The helium synthesis that fuels stars such as the Sun. The pale spheres represent protons, and the dark spheres neutrons. The other symbols are ν, neutrino; γ, gamma-ray photon; e⁺, positron.

The synthesis of helium-4 is responsible for the vast majority of the energy produced by the Sun. Every second, 655 million tonnes of hydrogen is converted into 650 million tonnes of helium in the Sun, and the five-million-tonne difference in mass is released as the vast energy output of the Sun. However, the Sun's bulk is so enormous that the mass lost in this way has no significant effect on the Sun's other properties, such as its gravitational pull on the Earth. We have seen that the amount of mass converted into energy when helium is produced is just 0.7 per cent of the total, so even if all the hydrogen in the Sun were converted to helium, the Sun would lose less than 0.7 per cent of its total mass.

The critical step in the fusion process is the initial one in which two protons are converted into a nucleus of heavy hydrogen, containing a proton and a neutron. The incredibly slow pace of this reaction determines the overall rate of the conversion of hydrogen into helium. It is fortunate for us that the weak force is so weak and that this reaction takes such a long time, because it means that the Sun will continue generating heat by converting hydrogen into helium for about ten billion years. At the present time the Sun is almost halfway through this stage of its life cycle. If the weak force had been much stronger, there would not have been enough time for complex organisms to have evolved on Earth. Nuclear fusion provides us with an energy source for the Sun that was unknown to Victorian scientists such as Helmholtz and Kelvin and explains how it is possible for the Sun to shine for billions of years: far longer than could have been imagined in the nineteenth century.

Celestial Alchemy

'Listen here, you thief with a degree, we possess a piece of information worth more than any other, a formula to fashion gold from ordinary atoms – for instance hydrogen, of which the Universe has an inexhaustible supply. We'll let you have it if you let us go.'

'I have a whole trunk full of such recipes,' answered the face, batting its eyes ferociously. 'And they're all worthless. I don't intend to be tricked again – you demonstrate it first.'

'Sure, why not? Do you have a jug?' 'No.'

'That's all right, we can do without one,' said Trurl. 'The method is simplicity itself: take as many atoms of hydrogen as the weight of an atom of gold, namely one hundred and ninety-six; first you shell the electrons, then knead the protons, working the nuclear batter till the mesons appear, and now sprinkle your electrons all around, and voila, there's the gold. Watch!'

And Trurl began to catch atoms, peeling their electrons and mixing their protons with such nimble speed, that his fingers were a blur, and he stirred the subatomic dough, stuck all the electrons back in, then on to the next molecule. In less than five minutes he was holding a nugget of the purest gold, which he presented to the face ...

Stanisław Lem, 'The Sixth Sally' (1975)[19]

The key attribute that determines the properties of a star and its life cycle is its mass. There is sufficient hydrogen in the Sun for the nuclear fusion reactions to continue, steadily converting hydrogen into helium, for about ten billion years. Bigger stars race through their life cycle much quicker. With stars it is certainly the case that the bigger they are, the faster they burn. The reason for this is that gravitational attraction increases with increasing mass, producing higher pressures and much greater temperatures in the core, where the nuclear fusion reactions take place. For instance, a star that is over twice as massive as the Sun, such as the brilliant bluish Vega, seen in the northern summer night sky, will convert all its hydrogen into helium in about a billion years. Over twice as much fuel is consumed in a tenth of the time.[20] A star that is fifteen times as massive as the Sun will deplete its hydrogen fuel in just over ten million years, a cosmic blink of the eye. Such a star will convert hydrogen to helium in its core at least 15,000 times faster than in the Sun. You can see a prominent example of such a star shining brilliantly in northern winter skies. This is the bluish-white giant Rigel, which forms one of the feet of the constellation of Orion. Although Rigel is over 800 light years away (200 times as far as the nearest star other than the Sun), it is so luminous that even at this distance it is one of the brightest stars in the night sky. Its location is shown on the star map in Figure 43.

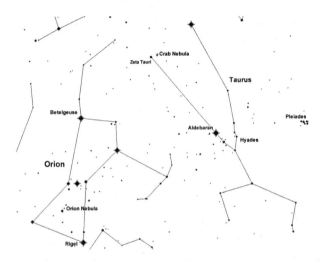

Figure 43 Star chart produced with *SkyMap Pro 11*, showing the constellations Orion and Taurus, with bright stars Betelgeuse and Rigel, the Crab Nebula and the Orion Nebula.

Fred Hoyle was one of the most imaginative physicists of the twentieth century.[21] His greatest achievement was to explain how the atoms that form the elements beyond number 2, helium were created. Hoyle was one of the most famous scientists of his era. In addition to being one of the world's foremost astrophysicists, he was one of the leading popular science writers and broadcasters. He also wrote best-selling science fiction novels, as well as a captivating BBC television serial called *A for Andromeda*. In his first novel, *The Black Cloud*, the solar system is visited by a gas cloud that turns out to be a highly intelligent sentient being of a nature totally different to terrestrial life.

Hoyle was raised in a small village close to Bradford in West Yorkshire. He was a no-nonsense Yorkshireman who was never afraid to speak his mind. Sometimes he could be blunt to the point of rudeness, and throughout his long career he never managed to stray far from controversy. He was also never afraid of entertaining apparently outlandish ideas. Some of his conjectures have turned out to be wrong, and others are still being debated, but most importantly, some of his ideas have proved to be spectacularly correct. Hoyle was the first to recognise that the rare, but exceptionally massive stars could be the sites where the elements heavier than helium are forged. He and three colleagues formed a team of nuclear alchemists who worked out the precise details. Astronomers refer to their groundbreaking work as B^2FH. These are the initials of the husband and wife team Geoffrey and Margaret Burbage, Willy Fowler and Fred Hoyle. Together they published a monumental paper in 1957 in which they described the nuclear reactions taking place in the biggest stars that would manufacture elements heavier than helium.[22]

The copious energy generated by nuclear fusion reactions in the core of a star provides an outward pressure that resists the collapse of the star under its own gravity. The Sun is a prime example of how this balance can be maintained for billions of years. However, eventually the star will exhaust its huge reserves of hydrogen. At this point the nuclear reactions in the core will no longer generate sufficient energy to prevent the collapse of the star. As the core collapses, the outer layers of the star expand until the star swells into a vast red giant. A star such as the Sun will eventually pass through this red giant phase. Its core will collapse into an extremely dense and hot body known as a white dwarf, while its outer regions disperse into space. The white dwarf star that remains after the red giant's outer layers have dispersed is effectively the glowing ember of the star's core. It is incredibly dense, with most of the mass of the star compressed into a sphere the size of the Earth. Nuclear fusion reactions have ceased in the white dwarf and the star will gradually cool as it radiates its heat into space.

In a star similar to the Sun, nuclear synthesis cannot proceed much beyond the element helium. Such a star will eventually manufacture carbon, nitrogen and oxygen, but nothing more. However, the B^2FH paper showed how exceptionally massive stars could synthesise much heavier elements, and it is the fate of these stellar prodigies that will concern us. When a heavyweight star has consumed its hydrogen fuel, it will undergo gravitational collapse until its core becomes sufficiently hot and dense for a new phase of nuclear fusion reactions to be initiated. Only in the more massive stars will conditions in the collapsed core become extreme enough for this to happen. The energy generated by the next stage of fusion reactions prevents further collapse of the star. Meanwhile, the outer layers of the star will have expanded enormously, so that the star becomes a bloated red supergiant, with a diameter as large as the orbit of Jupiter around the Sun. This will be the fate of the blue giant Rigel within the next few million years. The prominent red star Betelgeuse[23] that forms one of Orion's shoulders has a mass nearly twenty times the Sun's and has already undergone this transformation. It has used up all of its hydrogen fuel and its outer envelope has swollen to enormous dimensions, while its dense core is undergoing a new round of nuclear fusion.

The position of Betelgeuse in the night sky is indicated on the star map in Figure 43.

Helium nuclei are very stable nuclear nuggets, which is why so much energy is released in their synthesis in stars and why they are emitted as alpha particles from radioactive heavy elements. But their stability makes further fusion reactions difficult to achieve. Combining two helium nuclei to forge element number 4, beryllium, proves to be impossible. The beryllium-8 nucleus that would result immediately falls apart again into two helium-4 nuclei. (An additional neutron is required to produce a stable beryllium nucleus, beryllium-9. But the extra neutron is readily released from this nucleus, and this was key to the role played by beryllium in the discovery of the neutron.)

To take the next step through the periodic table beyond helium, it is necessary for three helium nuclei to collide simultaneously to form a nucleus of element number 6, carbon-12. Such triple collisions are very unlikely events. They are possible only if the helium nuclei are slammed together at extremely high temperatures, of around 60 million degrees Kelvin – much higher than the temperature at the core of the Sun. These temperatures are reached only in very massive stars such as Betelgeuse that have already consumed their hydrogen fuel. This reaction can be represented as

$$^4\text{He} + {}^4\text{He} + {}^4\text{He} \rightarrow {}^{12}\text{C} + \gamma + 7.4\,\text{MeV}$$

At these temperatures the carbon-12 will also undergo the next step in nuclear burning and fuse with another helium-4 nucleus to produce oxygen-16, an isotope of element number 8, oxygen. The carbon and oxygen produced in this way, being heavier than helium, will sink to the centre of the star, and in a metaphorical sense form the 'ash' left over from helium nuclear fusion. At the centre of red supergiants such as Betelgeuse, carbon and oxygen – two of the most important elements for the existence of life – are simultaneously being synthesised in vast quantities.

Helium burning generates less energy than the burning of hydrogen to form helium, so this stage of the star's life is correspondingly shorter. 7.4 MeV is released for each carbon-12 nucleus that is synthesised, compared with the 26 MeV released in the production of each of the three helium nuclei from which the carbon nucleus is formed. Altogether, $3 \times 26 = 78$ MeV is generated when the three helium nuclei were originally produced, and that is ten times the energy released when they fuse to form carbon. For this reason the fusion of helium to form carbon will continue for about a tenth of the time it took for the helium to be forged from hydrogen. In a star like the Sun, the helium-burning phase will last for a tenth of its total life span, about a billion years. But in a huge, fast-burning star of about twenty solar masses, like Betelgeuse, one which has already consumed most of its hydrogen and entered the helium burning phase, it will take about a million years for the helium to be exhausted. When the helium fuel has been depleted, the star's core will undergo further collapse and the temperature and pressure will rise again until the next stage of fusion is triggered. When the temperature has reached around 100 million degrees Kelvin and the density has approached one tonne per cubic centimetre, with a hop and a skip we can proceed another two places through the periodic table. At this temperature and pressure oxygen nuclei and helium nuclei are smashed together and will fuse to form element number 10, neon.

Neon nuclei will then fuse with helium nuclei to form element number 12, magnesium, the element that is vital for the collection of the Sun's rays by plants, the starting point of almost all the world's food chains.

The heavier nuclei will continue to fall towards the centre of the star, so that the star's core becomes stratified in a succession of shells in which different nuclear fusion processes are taking place, as shown in Figure 44. In Betelgeuse, the

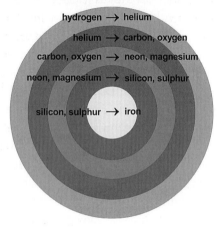

Figure 44 Nuclear fusion reactions that take place in concentric shells of a star on the verge of exhausting its nuclear fuel and collapsing into a neutron star. Hydrogen is element number 1, helium is element 2, carbon is element 6, oxygen is element 8, neon is element 10, magnesium is element 12, silicon is element 14, sulphur is element 16 and iron is element 26.

burning of carbon and oxygen to form neon and magnesium will continue for about a hundred thousand years. When all the carbon has been consumed the collapse will continue, with an inexorable rise in temperature and pressure, until further fusion reactions produce element number 14, silicon, and element number 16, sulphur. This stage is expected to last a mere twenty years. By the end of this period a wide range of the lighter elements are being cooked up in a complex broth involving many different nuclear reactions. The groundbreaking B²FH paper spelt out how all these elements could be forged in the dense heart of a massive star.[24] But the crucial question is: how did all these atoms come to find themselves within our bodies?

There is a natural endpoint for these stellar nuclear reactions. The nucleus of element number 26, iron, is the densest and therefore the most stable of all atomic nuclei.[25] This means that no further energy will be released in nuclear reactions involving the iron nucleus, and so, once this stage has been reached, the star cannot generate any further energy by nuclear fusion. And without the continued release of energy, the outward pressure that is preventing the gravitational collapse of the star cannot be maintained. The consequences of turning off the nuclear fusion reactor in the star's core are dramatic.

The Crab Nebula

> In the fifth month of the first year of the Chih-Ho reign period, Yang Wei-Tê, the Court Astronomer, announced, 'Prostrating myself, I humbly observe that a guest-star has appeared with an iridescent yellow glow. Respectfully, if one carefully examines the prognostications in accordance with the disposition for emperors, the interpretation says, "The guest-star does not infringe upon 'Bi', the Celestial Net. This shows that a Bountiful One is Lord, and that the country has a Great Eminence." I beg that this prognostication be handed over to the Bureau of Historiography to be preserved.' All

the Officials offered their congratulations and the Emperor
ordered that it be recorded.

Sung Hui Yao ('Essentials of the Sung Dynasty', 1054)[26]

Close to one of the horns of the constellation of Taurus, the
Bull, lies a fuzzy patch of light. This nebula is too faint to
be seen with the naked eye, but it is one of the most widely
studied astronomical objects in the night sky. A drawing of
it made in the 1840s by the Earl of Rosse, using his giant
telescope at Birr Castle in Ireland, was reminiscent of a
horseshoe crab, and the nebula has been known as the Crab
ever since. The Crab Nebula is about 6,500 light years away.
It is an expanding region of gas that is currently about 11
light years across. Analysis of the rate at which the material
that forms the nebula is moving outwards indicates that the
expansion has been under way for around 950 years.

The Crab Nebula lies about one degree – or twice the
apparent size of the Moon's disc – from the star that forms
the southern horn of Taurus. In clear, dark winter skies
the Crab Nebula can be seen in a small telescope, or even
glimpsed through binoculars in extremely good viewing
conditions. To find it, first find the familiar V-shaped head
of the bull formed by the Hyades cluster of stars, referred to
as the 'celestial net' in the Chinese records, and the bright
orange star Aldebaran[27] (see Figure 43). The arms of the
V point towards the two stars that form the horns of the
bull. Through a small telescope at low magnification, the
Crab Nebula appears as a faint oval smudge of nebulos-
ity, in the same field of view as the star Zeta Tauri, which
forms the southern horn of the bull. The position of the
Crab Nebula in the sky coupled with its rate of expansion
provide conclusive proof that it should be identified with

the 'guest-star' whose dazzling presence was recorded by the Chinese court astronomer in the ancient annals almost a millennium ago, as described in the preceding quote.

In 1968, a very powerful pulsating radio source was discovered within the Crab Nebula. This object beams radio waves in our direction about thirty times every second with incredible regularity, like a cosmic lighthouse. It is known as a pulsar,[28] and it is produced by the prodigiously dense compact remnant of the star whose supernova explosion was observed by the Chinese in 1054.

A supernova fireball is triggered when a supergiant star has finally consumed all its nuclear fuel. With its core transformed into iron, nuclear fusion reactions can proceed no further and no more energy is generated to withstand the inward gravitational pressure of the star's enormous mass. At this point the star's core collapses in an instant. Within one second a huge mass, about twice that of the Sun, is squeezed into a ball with a diameter of a mere 30 kilometres, about the size of a major city such as London.

In the final milliseconds of collapse, protons and electrons are forced together until they interact via the weak force in a process that is the opposite of beta decay. The electrons and protons merge into neutrons, and neutrinos are emitted:

$$e^- + p \rightarrow n + \gamma$$

The result is that the core of the star is transformed into a body that is composed almost entirely of neutrons. This is the extraordinary object that will remain after the outer layers of the star have been blasted deep into space in the supernova inferno. It is called a neutron star. It is just such a star that is responsible for the pulsar at the heart of the

Crab Nebula. As a neutron star forms, a colossal flux of neutrinos is emitted, one for each proton in the core that has been transformed into a neutron. The resulting neutron star has about twice the mass of the Sun. Its density is comparable to the density of an atomic nucleus – the star has, in effect, turned itself into a giant atomic nucleus.[29] Just one teaspoonful of neutron star material weighs about two billion tonnes.[30] At least, this would be the weight of an average teaspoonful of neutron star. Neutron stars are compressed to much higher densities towards their centres than at their crusts, so the weight of a teaspoonful of material scooped out from the core would weigh even more. The humble weak force is responsible for this remarkable transformation of the matter that formed the original star into the bizarre object that is the neutron star.[31]

The gravitational collapse of the bulk of the star into a tiny volume releases an enormous amount of gravitational energy. If you pick up a heavy weight and drop it, it is clear that when it hits the ground a lot of energy is released – especially if it lands on your foot. When the mass of the star's outer layers falls down and lands on the tiny neutron star forming at its core, the release of gravitational energy is stupendous. It has around the same magnitude as the total energy output of the star from nuclear fusion reactions throughout its entire life span. This is a truly gargantuan release of energy. Our galaxy is home to hundreds of billions of stars, but for several months after the initial explosion a supernova will shine with the brilliance of the entire galaxy. However, the visible light and electromagnetic radiation of other wavelengths that blaze out of the supernova represent only about 1 per cent of the total energy that is emitted. Remarkably, the rest of the energy is emitted in the form of neutrinos. The ultimate source of this vast power output is the energy released by the gravitational collapse of the star.

The sudden collapse of the core momentarily squeezes the massive neutron ball beyond nuclear densities. The stiff nuclear material then rebounds, sending a shockwave outwards. This shockwave travels outwards until it ploughs into the outer layers of the star that are now beginning to fall towards the collapsed core. The energy in the shockwave starts to falter as it encounters this mass of infalling material and begins to dissipate just as an enormous blast of neutrinos reignites the explosion.

By this point, the temperature of the collapsed core has risen to around 100 billion degrees Kelvin. At these enormous temperatures, gamma rays are emitted with sufficient energy to convert into electron–positron pairs. These electrons and positrons then annihilate to produce more gamma-ray photons and neutrinos. It is the outflux of these neutrinos that enables some of the energy to escape from the core and allows the exploding material to begin to cool. This vast outflow of neutrinos – about ten for every neutron in the neutron star, a total of 10^{58} (one followed by 58 zeros) – carries away about 99 per cent of the energy produced in the supernova explosion. These copious quantities of neutrinos race outwards from the core towards the stalling shockwave. Although neutrino interactions are extremely rare, the matter is so dense and the quantity of neutrinos emitted so mind-bogglingly enormous that the neutrino blast is sufficient to transfer huge amounts of energy and momentum into the outflowing matter and reignite the explosion. According to current theories, backed up by computer simulations, without this vast neutrino flux the supernova would turn into a gigantic damp squib. Who would have imagined that the humble weak force and the elusive neutrino would turn out to be responsible for the biggest fireworks in the cosmos?

The energy of the supernova blasts several solar masses of material into interstellar space. This is the origin of the matter that we see as the Crab Nebula. Eventually, this material will mix with the matter ejected from many other supernova explosions to form the gas clouds and dust lanes in the Milky Way that dim our view of much of the galaxy.

Ashes to Ashes, Dust to Dust

Hoyle and Fowler realised in 1960 that the energy released in a supernova explosion offered a natural solution to the problem of how the elements beyond iron were synthesised. These vast conflagrations of highly energetic atomic nuclei bathed in high-energy neutrons and protons were the perfect nuclear stews for cooking up the rest of the elements. These include the radioactive elements, such as uranium, which effectively store some of the energy of the supernova to release later within the Earth, thus keeping the Earth from freezing over and confounding Lord Kelvin's calculations. As Hoyle and Fowler proposed, the tremendous shock wave that blew the outer region of the star apart in a huge fireball would provide the enormous temperatures and fantastic release of energy that were necessary to generate a whole periodic table of elements and disperse them into interstellar space. This is the material from which new star systems condense. The solar system was born in the dust clouds formed from the wreckage of earlier generations of stars that had exploded as supernovae and spread their valuable cocktails of heavy elements throughout the galaxy.[32]

Plate 1 Computer-generated image with (above) snowflake-like symmetry produced with a virtual two-mirror kaleidoscope and (below) the symmetry of a tessellation of equilateral triangles produced with a virtual three-mirror kaleidoscope.

Plate 2 Computer-generated image with icosahedral symmetry produced in a virtual kaleidoscope.

Plate 3 Computer-generated image of a honeycomb of cuboctahedra and octahedra produced in a virtual kaleidoscope.[i]

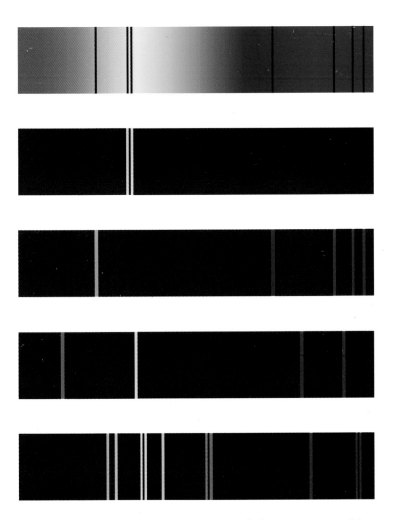

Plate 4 A schematic solar emission spectrum and absorption spectra of four elements. From top to bottom: emission spectrum of the Sun, and absorption spectra of sodium, hydrogen, lithium and mercury. The dark lines in the solar spectrum correspond to the bright lines in the spectra of sodium and hydrogen, but not lithium or mercury, demonstrating that the Sun's solar atmosphere contains sodium and hydrogen, but not lithium or mercury.

Plate 5 The Brocken spectre[ii]

Plate 6 The hero of H.G. Wells's *The World Set Free* was inspired by seeing the radioactive decay of individual atoms in a spinthariscope – an instrument invented by the physicist William Crookes. The Atomic Energy Lab shown in the illustration was sold in the early 1950s and included a spinthariscope, so that children could 'see actual atomic disintegration'. It also included a Geiger–Muller counter, and hinted at a '$10,000 Govt. bonus' for discovering deposits of uranium-containing ores.

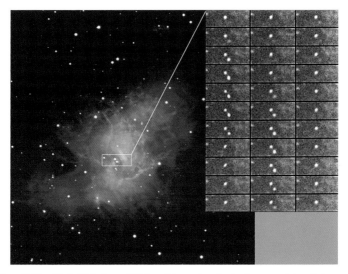

Plate 7 The Crab Nebula and Pulsar. The pulsar is the bottom right of the two stars in the centre of the nebula. The pulsations of the pulsar can be seen in the inset sequence of photographs of the central region of the nebula, which is a mosaic of thirty-three time slices, each of which represents approximately one millisecond.[iii]

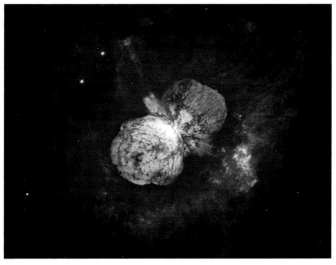

Plate 8 The hypergiant star Eta Carinae is surrounded by billowing dust clouds that it has ejected in its periodic eruptions. (Eta Carinae is a binary star, with one component of around 100 solar masses and the other about 30.)

Plate 9 Peter Higgs photographed on his eightieth birthday, 30 May 2008, in a lecture room at the James Clerk Maxwell Building, University of Edinburgh. Notice the Mexican hat diagram on the left of the blackboard.

Plate 10 Aerial flight – the Main Ring and Main Injector at Fermilab. The Main Ring is the circuit in the background; in the foreground is the Main Injector, completed in 1999.

Plate 11 'Moonrise over the Highrise', a photograph by Reidar Hahn, showing Fermilab's Wilson Hall and, to its left, Wilson's statue 'The Hyperbolic Obelisk'.

Plate 12 Inside ATLAS before the detectors were installed, showing the scale of the machine. The eight large tubes are the superconducting magnets that bend the trajectories of the newly created particles within the detector. When in operation, the proton beams are fired straight through the centre of this photograph and intersect at the heart of the detector.[iv]

Plate 13 Mural on the building above the ATLAS detector at CERN by Josef Kristofoletti. It depicts a cutaway diagram of ATLAS and a collision event within the detector that has produced a Higgs, which has then decayed into a muon–antimuon pair and an electron–positron pair. (This is the decay depicted in the centre of Figure 78.)

We can still see this process happening today. In the constellation Orion there is a dusky patch of light, easily seen with the naked eye in a dark clear sky, that forms the middle part of Orion's sword (see Figure 43, page 223). This is the Orion Nebula, an object that looks spectacular through a telescope. The Orion Nebula is illuminated by numerous bright young stars that have condensed within the last million years or so from the dust clouds that form the nebula. New stars are still forming in the nebula. Around four and a half billion years ago our Sun coalesced from just such a cloud, composed of the debris of earlier supernova explosions and containing the elements vital for a complex life-promoting environment like the Earth. It is quite remarkable that scientists have shown exactly how the elements are formed within the stars and also how we come to find them here in our environment.

This gives us a new perspective on the poetic Biblical account of the creation of humankind. The English burial service captures the cycle of life beautifully in the phrase 'ashes to ashes, dust to dust', which is an adaptation of a line from the Book of Genesis. We may not have been moulded from the clay of the Euphrates delta, but our flesh and bones are both the ash from the nuclear furnace at the core of a mighty star, and the stardust that was dispersed into the cosmos in an almighty supernova explosion.

The Pulsar

The collapse of the star at the heart of the supernova greatly increases the rate of its rotation. The neutron star that remains will be spinning many times a second – thirty times a second in the case of the neutron star within the Crab

Nebula. This rapid rotation generates a magnetic field whose intensity is around a trillion times that of the magnetic field of the Earth. From each pole of this intense magnetic field a beam of radiation is blasted into space. If the magnetic field of the neutron star is not aligned with its axis of rotation, in the same way that the Earth's magnetic poles are not aligned with the Earth's axis, then these beams will sweep around the heavens like a cosmic lighthouse, warning intrepid cosmonauts navigating the oceans of space to steer well clear. It is the beam of radiation from the rapidly spinning neutron star at the centre of the Crab Nebula passing through our line of sight thirty times a second that creates the regular pulses of radiation we detect.

Plate 7 shows the Crab Nebula. To the right of the main photograph are a series of images of two stars in the heart of the nebula. The bottom right star of this pair is the neutron star left behind by the supernova explosion that created the nebula. Because the neutron star is so tiny, it is extremely faint. It disappears completely, except when the blast of its pulsar beam is fired in our direction twice every rotation. The rotation of a neutron star is so rapid that the flashes of the pulsar cannot be distinguished by the naked eye, but they are clear in this series of photographs, each of which corresponds to a time slice of just one millisecond.

On a clear night from a dark location we can see the Milky Way arching overhead. This pale band is the combined light of innumerable stars, too faint to be seen individually, that form the plane of our galaxy's disc. The vastness of the galaxy is difficult to comprehend. The diameter of the disc is an enormous 100,000 light years. We live about 30,000 light years from the galactic centre. Of the hundreds of billion of stars in our galaxy, it is only the biggest that will eventually undergo a supernova explosion. Maybe just one in a million is such a behemoth. There are probably just a few hundred thousand of these supergiants in the galaxy at any one time. To get an idea of what the star inside the Crab Nebula looked like just before its spectacular supernova explosion, take a look at the same region of the night sky where there is a great example on show. The brilliant red supergiant Betelgeuse that forms the left shoulder of the giant Orion has the features we would expect of a star that will undergo a massive supernova explosion in the not too distant future. But this could be any time in the next million years or so, so we are unlikely to be around to see it.

However, in the skies of the southern hemisphere there is a remarkable star that could be on the verge of blasting itself apart very soon indeed, possibly even within our own lifetimes. This star is called Eta Carinae (see Plate 8), and it is one of the biggest stars in the galaxy, with a mass of over a hundred Suns and several million times as bright. Even at a distance of around 8,000 light years, it is visible to the naked eye. In recent centuries its brightness has changed quite erratically. At times it has appeared quite faint or even below naked-eye visibility, but on other occasions it has flared up dramatically. In 1843, it was for a period the second brightest star in the night sky after Sirius, which is one of our nearest stellar neighbours, less than one-thousandth of the distance to Eta Carinae. This huge star is reminiscent of a belligerent ruddy-cheeked giant with serious indigestion, belching and roaring its fiery breath into the depths of space, a broiling cauldron of extremely hot, dense matter churning around in its innards. Perhaps its next eruption will be its last, and we will witness the dazzling display of a supernova explosion.

Our knowledge of the lifespan of heavyweight stars, which is confirmed by studying other galaxies, suggests that there should be a supernova explosion in our galaxy once every thirty years or so. However, a supernova event in our galaxy has not been seen since 1604, just a few years before the invention of the telescope. Johannes Kepler kept detailed records of this supernova. Most of the galaxy is hidden from our view by the clouds of interstellar dust ejected by earlier generations of supernovae. We have a clear view of less than one-eighth of the galactic disc, and this is why the majority of supernovae in our galaxy go unobserved. We therefore have to look elsewhere to study these spectacular outbursts. There are around a hundred billion galaxies in the visible universe. Every year hundreds of supernovae events are recorded in distant galaxies, and although some large observatories operate automated searches for these cosmic eruptions, amateur astronomers also make many discoveries.

In 1987 there appeared the closest supernova to be observed since the one seen by Kepler. On 23 February that year, the Kamiokande laboratory in Japan recorded an unprecedented pulse of neutrinos. Eleven neutrinos were snared by the detector. At two other sites, one in America and one in Russia, neutrino detectors recorded another thirteen neutrinos. All twenty-four events occurred within a mere 13 seconds. The incredible rarity of neutrino interactions implies that the Earth must have briefly been bathed in a colossal flux of these elusive particles. The age of neutrino astronomy had begun. These were the first neutrinos to be detected that had originated from outside the solar system. They had been generated in the collapse of a blue supergiant star with around twenty times the mass of the

Sun – about the same as Betelgeuse – that had blasted itself apart in a supernova explosion. The location of the supernova was the Large Magellanic Cloud, a small irregular galaxy that is orbiting our own Milky Way Galaxy and lies around 170,000 light years away. The twenty-four neutrinos detected on Earth were a minute proportion of an estimated ten billion that had passed through every square centimetre of the Earth's surface and of the bodies of every one of us on the planet. Even at such an immense distance, the supernova was easily visible to the naked eye. At its peak it was not quite as bright as the stars of the Plough, but it was comparable in brilliance to the stars of another prominent northern constellation, the W-shaped constellation Cassiopeia. Unfortunately for northern observers, the supernova was visible only from the southern hemisphere.

The detection of this handful of neutrinos that had travelled such an enormous distance through intergalactic space was a great achievement and a fantastic verification of our current models of stellar evolution and the synthesis of the chemical elements. It confirmed our understanding of the high-energy processes that take place at the end of a star's life in terms of elementary particles. The weak force has played a vital part in the making of our universe. It is responsible for brewing up the wonderful cocktail of elements that make our universe complex and habitable. These processes take place at extremely high temperatures in the cores of massive stars. The elements are then disseminated throughout the cosmos in the most powerful events in the universe – supernova explosions. So, without the weak force we certainly would not be here to gaze up at the stars. As we will see, the clue that enabled physicists to understand the true nature of the weak force came not from the stars, but from a weird quantum phenomenon that manifests itself at extremely low temperatures. First, we must take a look at the other force that operates in the subatomic domain – the strong force.

Chapter Six

ZEN AND THE ART OF QUARK DYNAMICS

While the baroque rules of chess could only have been
created by humans, the rules of Go are so elegant, organic,
and rigorously logical that if intelligent life forms exist
elsewhere in the universe, they almost certainly play Go.

Edward Lasker, chess International Master (attributed)

The Glass Bead Game

Originally devised in the distant past of ancient China,
the board game Go was introduced into Japan along with
Buddhism in the ninth century. And it is in Japan that
playing Go has been refined into an art form. In the
seventeenth century the Shogun established four great Go
academies that still flourish to this day. A ranking system

of nine grades or dans was introduced for professional Go players along the same lines as for the martial arts. There could only ever be a single player in the ninth dan at any one time, and this player had to be clearly superior to all other players. He was given the title Meijin. Despite the rivalry between the Go academies, playing Go is a ritual activity that immerses the players in a deep contemplation of the forces of the universe.

The Honinbo Academy has produced more Meijin than all the other academies combined. The last great Honinbo master, Shusai, ceded his title in 1938 when he lost his final game, which took an incredible six months to play. Shusai's final epic match is retold in a novel by Yasunari Kawabata with the Japanese title *Meijin*, translated into English as 'The Master of Go'. Kawabata had originally covered the contest as a correspondent for the newspaper that sponsored the tournament. Ever since this celebrated event, the Japanese Go title has been decided in an annual tournament.

Japan's leading players have spent a lifetime mastering the game's complexities. The Chinese commentator Zhang Yunqi listed the qualities that are required to succeed at Go: 'the tactic of the soldier, the exactness of the mathematician, the imagination of the artist, the inspiration of the poet, the calm of the philosopher, and the greatest intelligence'.[1] Much of the fascination of Go derives from the fact that, although the game exhibits tremendous complexity, the rules are extremely simple. Go is a game for two players, who alternately place one of their playing pieces, called stones, on the intersections of the 19×19 grid that forms the Go board. There are a total of 361 intersections on the 19×19 grid, and as black always plays first there are 181 black stones in a Go set and 180 white stones. The aim of the game is to capture the largest amount of territory on the board by completely surrounding it with stones of your own colour.

There are only three rules for playing Go:

1. Individual stones or groups of adjacent stones that are completely surrounded by enemy stones are captured and removed from the board.
2. Suicide moves are not allowed: a player cannot place a stone in a position in which it is already surrounded by the opponent's stones.
3. It is prohibited to play a stone in a place that would exactly repeat the pattern on the board from a previous move. This restriction prevents the endless repetition of a position, but also has very important tactical consequences in the game.

At the end of the game the score of each player is calculated by adding together the number of enemy stones they have captured and the number of intersections of the grid that are within the territory they have completely surrounded. Although the rules of Go are so simple, the play is very subtle. In the words of one of the former holders of the Honinbo title, the ninth-dan player Iwamoto Kaoru, 'Go uses the most elemental materials and concepts: line and circle, wood and stone, black and white, combining them with simple rules to generate subtle strategies and complex tactics that stagger the imagination.'[2]

When a game of Go has been completed, the final position on the board is a work of art in itself (see Figure 45). The black and white stones seem to represent the flow of opposing forces, like ripples in a stream or veins in a jade sculpture. Complicated patterns begin to emerge within a few moves of the opening. The ability to play well depends on being able

Figure 45 The final position in a game of Go. This image, titled 'Goban', was created by Juha Nieminen, using the POV-Ray ray-tracing program.[3]

to recognise familiar patterns of stones, but there is also an intuitive element to good play, and experienced players seem to feel their way around the board rather than calculating the outcome of every move. Success requires the global pattern of stones across the whole board to be read correctly, which makes it very difficult to quantify the value of any particular move, and for this reason, it has proved surprisingly hard to develop software that can play Go to a high standard. There are no Go programs that can play the game at a level that is comparable to computer chess programs. A competent, but not necessarily exceptional, Go player can comfortably defeat even the best computer programs. Go remains an activity where the human mind, communing with the complex forces of the universe, can still take on the processing might of a modern computer.

Go is a very good example of how extremely simple rules can provide a framework for very complex behaviour.

Although the rules governing each individual move are so simple, play becomes incredibly involved as the stones combine to form intricate patterns on the board. The 19×19 Go board is like a model of the cosmos in miniature and there are interesting parallels with the structure of the real world that we inhabit. The interactions of the matter that we see all around us lead to an overwhelming diversity of phenomena. This complexity and diversity is exhibited at every level in the universe. We can see it in the nuclear furnaces within massive stars, brewing a heady cocktail of chemical elements which are later dispersed throughout the galaxy in a supernova explosion. We see it in the evolution of complex life forms on our planet, and we see it again in two eminent Japanese masters immersed in the intellectual combat of Go. As this chapter will reveal, beneath the complex interactions visible on the surface, the rules of the game can be remarkably simple, and the same is true of the fundamental laws that determine the structure and the behaviour of the matter of which the universe is formed.

Divine Symmetry

We have already taken a close look at the electromagnetic and weak forces, but there is a third force that plays a vital role in nuclear physics. This is the force that binds together the atomic nucleus – the strong force. The distinction between those particles that feel the strong force and those that do not is fundamental to the composition of matter. Electrons do not feel the strong force. They remain outside the nucleus, interacting with the protons in the nucleus only via the electromagnetic force. Neutrinos feel neither the strong force nor the electromagnetic force, which is why they are rarely detected. But this chapter will be concerned with the particles, such as the proton and the neutron, that do feel the effects of the strong force. Such particles are collectively known as hadrons, and this is the origin of the name of the Large Hadron Collider.

The discovery of the neutron had enormous implications. Its existence immediately clarified the physics of the atomic nucleus. For good or ill, humanity now had a powerful insight into the structure of matter at a deeper level. It became possible to explain the power output of the Sun and stars, as we have seen, and it would take just a few years for the technology of nuclear power stations and nuclear weapons to be developed. The speed with which these possibilities became realities shows Chadwick's discovery to be the most significant of the many particle discoveries of the twentieth century. Looking back in *Scientific American* two decades later, Robert Marshak wrote:

> In the year 1932, when James Chadwick discovered the neutron, physics had a sunlit moment during which nature seemed to take on a beautiful simplicity. It appeared that the physical universe could be explained in terms of just three elementary particles – the electron, the proton, and the neutron. All the multitude of substances of which the universe is composed could be reduced to these three basic building materials, variously combined in 92 kinds of atoms.[4]

Theorists immediately began to grapple with the significance of the neutron. It was clear that the existence of a nucleus composed simply of protons and neutrons makes sense only if there is a force holding the positively charged protons and the uncharged neutrons together. This force must be much stronger than the electromagnetic repulsion between the protons, otherwise this repulsion would blast any nucleus apart in an instant. The glue that holds the nucleus together is so powerful that physicists call it the strong force. The next challenge was to explain how the strong force operates.

Werner Heisenberg, one of the pioneers of quantum mechanics, led the way. Few clues were available, as nuclear physics was in its infancy, but Heisenberg devised an argument based on symmetry. His approach demonstrates the power of symmetry for tackling physical problems: it often enables far-reaching conclusions to be drawn without the need for detailed knowledge of the processes that are involved. In Chapter 3 we saw how, if waves are confined to a symmetrical region, as in the 'square-drum' atom, they fall into sets with the same energy. Similarly, at much higher energies in particle physics, the equivalence of mass and energy means that particles fall neatly into sets with the same mass whenever they are related by some sort of symmetry. Conversely, the discovery of a collection of particles with the same mass is an indication of a new symmetry of nature.

Heisenberg began his assault on the strong force in 1932 with two very simple observations. He noted that the proton and the neutron have almost the same mass. Also, an atomic nucleus typically contains similar numbers of protons and neutrons – for instance, the nucleus of a helium atom (^{4}He) is composed of two protons and two neutrons, and the nucleus of the most abundant carbon atom (^{12}C) is composed of six protons and six neutrons. These two facts suggested to Heisenberg that the strong force affects protons and neutrons in exactly the same way. Clearly, this is in stark contrast to the electromagnetic force: the proton carries a positive electric charge, whereas the neutron carries no electric charge, so the proton feels the electromagnetic force but the neutron does not. Heisenberg realised that if we could turn off the electromagnetic force and just consider the strong force, the close similarity between the proton and the neutron would become apparent. As far as the strong force is concerned, the proton and the neutron are twins.

Heisenberg deduced that there must be a fundamental symmetry underlying the operation of the strong force. He reasoned that if the identity of all the protons and neutrons in the universe were suddenly swapped, so that all the neutrons were transformed into protons and all the protons were transformed into neutrons, then nothing would change as far as the strong force was concerned. He called this symmetry 'isospin'. (This is perhaps an unfortunate name as isospin has nothing to do with spin, although it is described using similar mathematics.)

Heisenberg now suggested that isospin symmetry implied the existence of a new type of charge – isospin charge – that would be conserved by the strong force. He reasoned that, at least with regards to the strong force, the proton and neutron could be considered as two faces of a single particle: the nucleon. He described them as an isospin doublet. The proton was a nucleon with $+\frac{1}{2}$ a unit of isospin charge, and the neutron was a nucleon with $-\frac{1}{2}$ a unit of isospin charge. The conservation of isospin charge is completely analogous to the conservation of electric charge. The sum of the electric charges before an interaction equals the sum of the electric charges after the interaction. But with isospin, the charge is conserved only in interactions produced by the strong force. Heisenberg used isospin symmetry to make predictions about nuclear interactions, and when these predictions were tested in the laboratory he was shown to be absolutely correct. It was a clear display of the power of symmetry as a tool in particle physics.

The identification of conserved quantities enables physicists to determine which processes can occur and which are ruled out, even without a detailed understanding of the forces. This approach, pioneered by Heisenberg, became the model for how the problems of particle physics should be tackled, and the search for conserved charges became a major goal for the researchers of the 1950s and 1960s. Each identification of a conserved quantity provided conclusive evidence for the existence of a symmetry of nature.

Although Heisenberg recognised isospin as an abstract symmetry of the strong force, he didn't offer any suggestion about its origin. Today we know the reason for this symmetry, and we can explain why the proton and the neutron are close siblings. The answer will soon be revealed.

The Strange New Particle Garden

The discovery of the neutron had produced a compelling picture of the atom that completed the explanation of the periodic table of the elements. The nucleus of an atom was composed of a collection of protons and neutrons, and this nucleus was surrounded by a swarm of orbiting electrons. But this simple pre-war picture of the structure of the atom was not to last. With the resumption of pure research after the Second World War, it soon became apparent that particle physics was much more complicated than had previously been imagined. The year 1947, exactly fifty years after J.J. Thomson's identification of the first subatomic particle, the electron, saw the beginning of an explosion of new particle discoveries. While Schwinger, Feynman and Dyson were unravelling the equations of QED, major experimental advances were being made on the other side of the Atlantic.

British physicists led by Donald Perkins, Cecil Powell and Patrick Blackett turned their attention back to cosmic rays, and photographic laboratories in the UK developed extremely sensitive emulsions specifically for their research. Teams of physicists from the universities of Bristol and Manchester took photographic plates to high altitudes in RAF aircraft and to peaks in the Alps and the Andes. Powell's team from Bristol made the first important post-war cosmic-ray discovery – a particle we know today as the pion. Pions are continuously being produced high in the Earth's atmosphere as high-energy protons from outer space slam into protons or neutrons in atomic nuclei in the air. Many years later, Powell reflected on the excitement he felt when he first examined the freshly developed photographic plates. Laid out before him was 'a whole new world. It was as if, suddenly, we had broken into a walled orchard, where protected trees flourished and all kinds of exotic fruits had ripened in great profusion.'[5]

It would take several years to arrive at a complete under-standing of this new terrain, and fully map out its intricate pathways, but exploration of this new world now became the top priority in particle physics. The first of the new particles, the pion, has a mass of about 140 MeV, which is about 280 times the mass of an electron and one-seventh of the mass of a proton. The fact that this mass was midway between the mass of the electron and that of the proton led to the pions being classed as *mesons*, which simply meant 'middle-weight particles'. There are three varieties of pion, distinguished by their electric charge: the negatively charged pi-minus (π^-), the neutral pi-nought (π^0) and the positively charge pi-plus

(π^+). Like protons and neutrons, pions interact via the strong force, and the triplet of pions fits quite naturally into Heisenberg's isospin scheme; the pi-plus (π^+) has an isospin charge of +1, the pi-nought has zero isospin charge and the pi-minus has an isospin charge of −1.

Towards the end of 1947, George Rochester and Clifford Butler at Manchester University announced that they had found two examples of another new particle in their cosmic-ray photographs, this time with around half the mass of the proton. The tracks produced by the decay of these particles into two lighter particles had a distinctive V shape in their photographs, so they referred to them as V particles. Their report was received with quiet scepticism. The identification of a new type of particle was a bold claim to make on the evidence of just two examples, but Rochester and Butler were convinced that their interpretation of the data was correct. Unfortunately they were unable to produce any further examples of the strange new particles they had found.

For the next couple of years, Rochester and Butler continued to present their findings to meetings of physicists, but their audiences remained unenthusiastic, and their inability to reproduce their results was becoming embarrassing. Eventually, while in the Californian city of Pasadena, Rochester met someone who took the implications of the unusual tracks in the photographs seriously. This was Carl Anderson, the discoverer of the positron. Anderson immediately arranged for his best cloud chamber to be taken to the top of White Mountain, the highest peak in California. The greater the altitude, the greater the intensity of the cosmic radiation, as there is a thinner blanket of atmosphere that the radiation must penetrate. If the V particles were real, they would certainly show up in Anderson's cloud chamber. To Rochester's great relief he soon received a letter from Anderson to say that thirty of the V particles had left their characteristic tracks in his cloud-chamber photographs.[6]

Over the next few years, teams of particle physicists would spend much of their time cooped up at high altitude attempting to decipher the filigree patterns in their photographs. A whole new catalogue of particles were about to make their debut. Despite the heroic efforts of the physicists to account for the new particles raining down from the heavens, their picture of the structure of matter was becoming ever more confusing. But beneath all the complexity, a new order would eventually be revealed. The leading architect of this new understanding of the structure of matter would be a young physicist from New York called Murray Gell-Mann. A child prodigy who had entered Yale University in 1945 at the age of just fifteen, Gell-Mann was a great lover of the diversity of nature, with an extremely wide range of passions from bird-watching to archaeology, linguistics and the arts. One aspect of Gell-Mann's creativity was his invention of catchy terminology that would stamp his indelible mark on the subject. For a twenty-year period from the early 1950s, he led the way in elucidating the structure of matter and the mechanism behind the strong force that holds the nucleus together.

Strangeness

> Oh I could rack my brain trying to explain
> Where it is I think that we are heading
> Strangely Strange, but Oddly Normal.
>
> Dr Strangely Strange, Strangely Strange But Oddly
> Normal (1969)

It is very difficult to interpret the interactions produced by showers of cosmic rays, as the precise energy and the exact identity of the particles cascading down on us from the depths of space are impossible to determine. Further progress in particle physics would require a return to the controlled environment of the laboratory. The 1950s saw the arrival of a new generation of circular particle accelerators, called synchrotrons, which propelled physics to ever higher energies. In 1953, Brookhaven National Laboratory in New York commissioned a synchrotron for their researchers, and this sexy, space-age machine was given a suitably futuristic name – the Cosmotron. With a design energy of 3.3 GeV, it was the first machine that could accelerate particles to energies in the billion electronvolt range before blasting them into a target. The paths traced out by the particle debris resulting from collisions could be photographed and analysed for signs of new physics.

The commissioning of similar accelerators for laboratories around the world inaugurated a new era in particle physics. The focus of research now descended from the mountain peaks to the labs, and soon new particles were being found at breakneck speed. The synchrotrons produced beams of particles whose energy could be precisely controlled, allowing the nature of the mysterious V particles to be investigated. It turned out that there were several different particles producing V-shaped tracks. These included particles of intermediate mass that were classed as new types of meson. They were heavier than the pions and were christened K mesons or kaons (pronounced 'kayons'). The kaons came in several varieties including an electrically neutral kaon, the K-nought (K^0), and a positively charged kaon, the K-plus (K^+).

It was soon realised that although the kaons were being created in interactions that happened almost instantaneously, with the characteristically short time span of the strong force, they decayed much more slowly. In fact, the process by which they were decaying was taking about ten trillion times as long as the process in which they were being produced. Independently, Murray Gell-Mann and a Japanese physicist, Kazuhiko Nishijima, explained this puzzling observation. They both proposed that the odd behaviour of the 'strange' new particles could be understood if they were carrying a new type of charge that is conserved in strong interactions, but not conserved in weak interactions. Gell-Mann, with his gift for a catchy name, called the new charge 'strangeness'. The idea was that when the particles were produced by a strong interaction between a pion and a proton, for instance, the total amount of strangeness would be conserved. For every new particle produced with strangeness +1, another particle would be produced with strangeness −1, so overall the sum of the strangeness would remain zero. And although each particle carrying the strangeness charge would be unstable, it would not be able to decay into particles with zero strangeness by the strong force, because the strong force conserves strangeness.[7] However, they could decay by the weak force, if the weak force did not conserve strangeness. This would explain why the life span of the strange particles was so long by nuclear standards. Their life span, one-tenth of a nanosecond, would be appropriate for the weak force.

The following reactions illustrate the creation and subsequent decay of kaons – in which strangeness is conserved as the kaons are created, but not conserved as they decay. Kaons are created via the strong force, extremely quickly:

$$\pi^+ + \pi^- \rightarrow K^+ + K^- + \pi^0$$
(strangeness) $\quad 0 + 0 \rightarrow +1 + -1 + 0$

The K-plus then decays, via the weak force, much more slowly:

$$K^+ \rightarrow \pi^+ + \pi^0$$
(strangeness) $\quad +1 \rightarrow 0 + 0$

The K-minus decays in a similar way.

The kaons were soon joined by an array of other 'strange' particles, including lambda particles (Λ^0), sigma particles ($\Sigma^-, \Sigma^0, \Sigma^+$) and cascade or xi particles (Ξ^-, Ξ^0). This was, at first sight, a bewildering collection of new particles, each with its own characteristic properties. But Gell-Mann realised that these properties were not completely haphazard, and that all the new 'strange' particles could be organised quite naturally into Heisenberg's isospin multiplets. Each multiplet consists of a set of particles with equal mass but different charges. And because isospin charge is conserved in strong interactions, the isospin charge of each particle could be cross checked by studying its interactions with other particles, such as protons, neutrons and pions, whose isospin charges had already been worked out. This enabled Gell-Mann to deduce that the sigma particles form an isospin triplet, with the sigma-minus (Σ^-) carrying isospin charge -1, the sigma-nought (Σ^0) carrying zero isospin charge and the sigma-plus (Σ^+) carrying isospin charge $+1$. Similarly, he determined that the two cascade particles (Ξ^-, Ξ^0) form an isospin doublet with isospin charges ($-\frac{1}{2}, +\frac{1}{2}$).

You may feel that this is all becoming rather complicated. If so, don't worry – most physicists would agree with you. Physicists don't like having to remember lots of names of different particles along with a confusing catalogue of their properties; they want to find elegant patterns with a lot of explanatory power. There were physicists in the 1950s who felt that their subject was becoming a bit like bird spotting or stamp collecting. Indeed, the great Enrico Fermi remarked to one of his young research students, 'Young man, if I could remember the names of all these particles, I would have been a botanist.'[8] And another leading physicist, Willis Lamb, began his Nobel Prize acceptance speech in 1955 by commenting that:

> When the Nobel Prizes were first awarded in 1901, physicists knew something of just two objects which are now called 'elementary' particles: the electron and the proton. A deluge of other 'elementary' particles appeared after 1930; neutron, neutrino, mu meson, pi meson, heavier mesons, and various hyperons. I have heard it said that the finder of a new elementary particle used to be rewarded by a Nobel Prize, but such a discovery now ought to be punished by a $10,000 fine.[9]

Fortunately, as we will soon see, there are very simple patterns that explain the particle profusion, just as in the board game Go, where the complicated interwoven threads of black and white stones result from very simple rules of play.

Attaining Nirvana

By the early years of the 1960s, the inventory of the subatomic world had expanded to include dozens of particles, and researchers were using the results of accelerator experiments to deduce the properties of the new particles that were being discovered. In particular, they worked out the values of the various charges that the particles carry, and then used these values to classify them. The proliferation of particles put particle physicists in a similar position to the chemists of a hundred years earlier. Surely, not all the new particles could have the same fundamental status – there had to be a pattern beneath the apparent disorder. By the 1860s, all chemical substances were known to be composed of about a hundred different elements; now history was about to repeat itself at the deeper level of nuclear interactions. with Murray Gell-Mann playing the role of Mendeleyev.

In 1961, Gell-Mann and the Israeli physicist Yuval Ne'eman both saw how isospin and strangeness could be combined to bring some clarity to the prevailing particle chaos. By considering the isospin and strangeness charges simultaneously, many particles could be arranged into patterns of octets. Gell-Mann romantically described the scheme as the Eightfold Way, taking the name from the Buddhist path to enlightenment. This was the crucial spark of inspiration that would eventually give rise to modern particle physics. In the philosophy of Buddhism, the Eightfold Way shows the path to inner contentment and mental development, aiming to free the individual from worldly attachments and delusions, ultimately bringing insight into the metaphysical truth about all things. It consists of right view, right intention, right speech, right action, right livelihood, right effort, right mindfulness and right concentration. But rather than being steps in a sequence, the eight principles of the path are equally important and must be considered together as a whole. To Gell-Mann this seemed an appropriate metaphor for the symmetrical organisation of the ethereal high-energy particles that would shed new light on the structure of matter at its deepest level.

Like a great Go master, Gell-Mann carefully placed his stones on the board and found a new fundamental pattern. The mysterious strong force was beginning to reveal its secrets. To see how Gell-Mann's scheme works, we can take a look at how some of the octets of his Eightfold Way fit together. The first thing to note is that the members of each isospin multiplet all carry the same strangeness charge. For instance, the two particles in Heisenberg's original doublet, the proton and the neutron, have zero strangeness, whereas the three sigma particles (Σ^-, Σ^0, Σ^+) that form an isospin triplet each have -1 unit of strangeness. And the isospin doublet of cascade particles (Ξ^-, Ξ^0) each carry -2 units of strangeness. It was a simple exercise for Gell-Mann to plot the isospin and strangeness charges of these particles, and when he did so they fitted together to form the pattern shown in Figure 46. Six of the particles form a hexagon, and the other two occupy the same point at the centre of the hexagon. This is the baryon octet. The baryons are the proton, the neutron and their relatives. The name means 'heavy particles', from the Greek *barys*, meaning heavy; though this term is still in use, its etymology no longer seems so appropriate, as we will see later. Ultimately, the reason why the particles fit together to form an octet, rather than say a septet or a nonet, is explained by the mathematics of group theory.[10]

Gell-Mann recognised the parallels with his nineteenth-century predecessor Mendeleyev, whose discoveries pointed the way towards understanding the structure of the atom and explaining chemistry in terms of atoms. Gell-Mann was now undertaking an equivalent task, aiming to bring order to the structure of nuclear matter. When Mendeleyev arranged the elements to form his periodic table, he realised that there were a number of gaps in the pattern he had found. But his

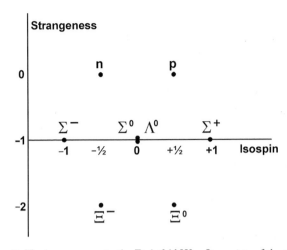

Figure 46 The baryon octet in the Eightfold Way. It consists of the two nucleons (n, p), three sigma particles ($\Sigma^-, \Sigma^0, \Sigma^+$), the lambda particle ($\Lambda^0$) and two cascade particles (Ξ^-, Ξ^0). The two particles shown at the centre of the octet, Σ^0 and Λ^0, both have strangeness −1 and zero isospin. One of these, the Σ^0 is part of an isospin triplet, and the other, the Λ^0, is an isospin singlet. The points representing these two particles are both situated at the centre of the hexagon. In this and the other octet diagrams in this chapter, the two points have been separated slightly in the diagram so that they can be distinguished.

confidence was not shaken. There was a profound beauty in his symmetrical arrangement of the elements, and he felt it had to be true. He boldly asserted that the gaps were not a flaw in his system, but a beacon which would guide us to new elements that would complete the table. So, he predicted the existence of several new elements and deduced their properties from his chemical patterns. The subsequent discovery of these hitherto unknown elements was a great triumph, and eventually all the empty spaces were filled. Almost a hundred years later, events would take the same course; the parallels would be almost uncanny.

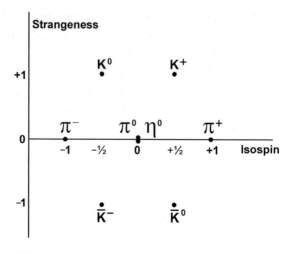

Figure 47 The meson octet in the Eightfold Way. There are two kaons (K^0, K^+), forming an isospin doublet with strangeness +1; three pions (π^-, π^0, π^+), forming an isospin triplet with strangeness 0; an eta particle (η^0), an isospin singlet with strangeness 0. The octet is completed by the two particles at the bottom of the diagram, forming an isospin doublet with strangeness −1, known as the K-minus-bar and the K-nought-bar. (A bar above a symbol generally indicates an antiparticle. Here, the K-minus-bar is the antiparticle of the K-plus, and the K-nought-bar is the antiparticle of the K-nought.)

Gell-Mann realised that by plotting the isospin and strangeness charges of the pions and kaons, he could form an octet of mesons, as shown in Figure 47. However, only three pions and four kaons were known, so there were only seven of the eight particles needed to form a full octet. But Gell-Mann was convinced that his Eightfold Way was the correct way to look at the symmetries of the strong force. He felt that the octet pattern of the mesons must be correct, so, with a self-belief comparable to Mendeleyev's, he predicted that a new particle would be found to fill the gap in the octet.[11] This hitherto unseen particle was named the eta particle, represented symbolically as η^0. Gell-Mann's dazzling insight into the symmetries of the natural world was to prove absolutely right. Before the end of 1961, accelerator experiments at the Brookhaven synchrotron had identified the eta particle and provided a sensational confirmation of his prediction in the laboratory.

Broken Symmetry

> Tyger! Tyger! burning bright
> In the forests of the night,
> What immortal hand or eye
> Dare frame thy fearful symmetry?
>
> William Blake, 'The Tyger' (1794)

As we are beginning to see, the goal of modern particle physicists is to reveal the fundamental symmetries of the universe. For isospin to be a true symmetry of the strong force, each of the particles related by the symmetry should have precisely the same mass. We can check whether this is the case for the nucleons – the neutron and the proton. The mass of the neutron is 939.6 MeV and the mass of the proton is 938.3 MeV. The difference is just 1.3 MeV, which is around 0.1 per cent of their mass, so the two particles do indeed have almost equal mass. This near equality means that although isospin is not a perfect symmetry of the strong force, it does hold to a very good approximation. This is, of course, why Heisenberg proposed isospin as a symmetry of the strong force in the first place.

Gell-Mann fitted together various other isospin multiplets to form the octets of the Eightfold Way. The success of his scheme indicated that the isospin symmetry was part of a larger symmetry of the strong force. It was clear, however, that the symmetry of the Eightfold Way is quite badly broken. If the symmetry were exact, then all the particles in an octet would have exactly the same mass. By looking at the masses of the particles in the octets we can see just how the mass of the particles is affected as strangeness decreases. The table below gives the mass and the strangeness of the particles that form one of Gell-Mann's octets. The mass of the particles in each isospin multiplet are so close together that only a rounded-up average for each isospin multiplet is quoted here. The difference in the masses of the isospin multiplets shows clearly that the Eightfold Way symmetry is broken. Decreasing the strangeness by one unit increases the mass of a particle by roughly 200 MeV, or about 20 per cent of the total mass.

Mass and strangeness values for the particles in the baryon octet.[12]

Particles	Mass, MeV	Strangeness
Neutron and proton (n, p)	940	0
Lambda particle (Λ^0)	1,115	−1
Sigma particles (Σ^-, Σ^0, Σ^+)	1,190	−1
Cascade particles (Ξ^-, Ξ^0)	1,320	−2

From Alpha to Omega-minus

Gell-Mann's name 'The Eightfold Way' was inspired by the patterns of particle octets that he assembled. But, the symmetry of the Eightfold Way does not only manifest itself in particle multiplets with eight members. Gell-Mann noticed that

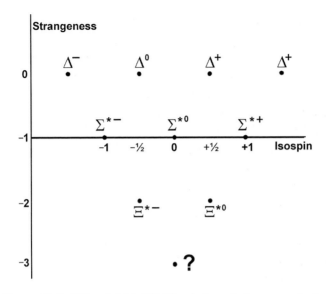

Figure 48 Gell-Mann's Eightfold Way decimet. It is composed of an isospin quartet of delta resonances (Δ), an isospin triplet of sigma-star resonances (Σ*) and an isospin doublet of xi-star resonances (Ξ*), with an empty space that Gell-Mann predicted would be filled by the omega-minus.

some newly discovered but extremely short-lived particles, known as resonances, fitted into a different type of multiplet. He called this triangular-shaped pattern of ten particles the 'decimet'. An isospin quartet of delta resonances (Δ), an isospin triplet of sigma-star resonances (Σ*) and an isospin doublet of xi-star resonances (Ξ*) gave Gell-Mann nine of the ten particles he needed to complete his decimet, as shown in Figure 48.

If Gell-Mann's analysis was correct, a tenth particle was missing from the bottom of the triangle. Here was another opportunity for Gell-Mann to predict the existence of a new particle. We can follow in his footsteps and deduce the properties of the hypothetical particle. In order to complete the decimet, it had to have zero isospin and a strangeness charge of −3. The electric charge of the particle could also be read from the diagram. All the particles on each diagonal from top left to bottom right have the same electric charge. For example, all four particles in the long diagonal on the left, including the missing particle, have a charge of −1. Gell-Mann named the absent particle the omega-minus, which is represented symbolically as Ω^-.

With its strangeness of −3, the decay of the omega-minus would have a clear signature. There are no lighter particles with three units of strangeness, so the omega-minus would be able to decay into lighter particles only by losing some of its strangeness. And as strangeness is conserved in strong interactions, the omega-minus would not be able to decay via the strong force. So, unlike the other particles in the decimet, which are all extremely short-lived, the omega-minus would decay by the weak force and would therefore have a lifetime of about a tenth of a nanosecond, characteristic of the weak force.[13] To give the experimentalists an idea where to look for signs of the new particle, Gell-Mann had to estimate the mass of the omega-minus.

Gell-Mann first presented his prediction of the omega-minus particle during an end-of-seminar discussion at a CERN conference in 1962. In the audience was Nick Samios, a young Greek researcher who was working at the Brookhaven National Laboratory. In the canteen after the seminar, Gell-Mann filled Samios in with a few more details about the identity of the new particle and scribbled down on a paper napkin how it might best be produced in an accelerator.[16] Bursting with enthusiasm for Gell-Mann's ideas, Samios returned to Brookhaven determined to track the new particle down. The

---

Puzzle 9

What is the mass of the omega-minus? If the masses of the delta resonances $(\Delta^-, \Delta^0, \Delta^+, \Delta^{++})$ are all around 1,236 MeV, the masses of the sigma-star resonances $(\Sigma^{*-}, \Sigma^{*0}, \Sigma^{*+})$ around 1,385 MeV, and the masses of the xi-star resonances, (Ξ^{*-}, Ξ^{*0}) around 1,530 MeV, what would you expect the mass of the omega-minus (Ω^-) to be?

---

hunt was on. By the early 1960s, Brookhaven had retaken the lead in producing the highest-energy particle beams in the world. Protons could now be accelerated to energies of up to 33 GeV before being smashed into a fixed target.[17] The trails of the particles produced in the debris from the high-energy impacts were then photographed in a chamber filled with liquid hydrogen at high pressure, known as a bubble chamber. For two years, Samios led his team in the quest for the omega-minus. The researchers analysed the intricate arabesques of pirouetting particles in almost 100,000 photographs before the unmistakable signature of the omega-minus was finally identified. The photograph in which the omega-minus was discovered in 1964 is shown on the left in Figure 49.

The magnetic field in the bubble chamber causes charged particles to curve: positively charged particles curve to the left in the photograph, and negatively charged particles curve to the right. Uncharged particles leave no visible trail; but they are not deflected by the magnetic field and so travel in a straight line through the bubble chamber, and this enables the paths they have followed to be worked out. At each interaction vertex the total momentum of all the particles involved in the interaction must be conserved. This allows the direction in which the uncharged particles are moving to

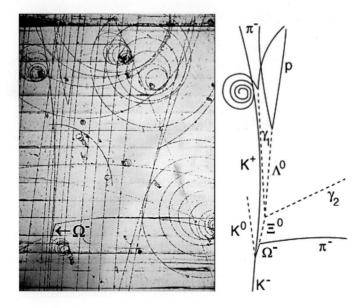

Figure 49 Left: The photograph taken in 1964 in which the omega-minus particle was discovered. Right: The identification of particles from their tracks in the photograph.

be deduced and their trajectories determined. To the right of the photograph in Figure 49 is a line drawing in which some of the tracks on the left side of the photograph have been identified. The uncharged particles that do not appear in the photograph are represented as dashed lines in the drawing. These trajectories have been determined by working backwards through the decay process, from the interactions at the top of the photograph downwards.

The lines in the drawing can be interpreted as follows. A negatively charged kaon (K^-) enters the bubble chamber from the bottom of the photograph. It collides with a proton in the bubble chamber and undergoes the following reaction:

$$K^- + p^+ \rightarrow \Omega^- + K^0 + K^+$$

The K-nought is uncharged and cannot be seen in the photograph; the K-plus bends towards the left and travels all the way through the photograph without interacting. The elusive omega-minus (Ω^-) travels a short distance before decaying into a neutral cascade particle (Ξ^0) and a negative pion (π^-):

$$\Omega^- \rightarrow \Xi^0 + \pi^-$$

Although the neutral cascade particle is not visible, the negative pion can be seen veering off to the right. The cascade particle then decays into a neutral lambda particle (Λ^0) and two photons (labelled γ_1 and γ_2):

$$\Xi^0 \rightarrow \Lambda^0 + \gamma + \gamma$$

None of these three particles is charged, so they do not leave a trail. One of the photons interacts with a proton towards the top of the photograph and produces an electron–positron pair. The positron forms the tight spiral to the left, and the electron curves away to the right. The lambda-nought (Λ^0) decays into a proton and a negative pion:

$$\Lambda^0 \rightarrow p^+ + \pi^-$$

The proton and the pion form a V shape towards the top of the photograph, with the proton track curving to the left and the pion track curving to the right. (The lambda-nought is one of the particles that were originally identified as V particles, because as it decays it produces trails that look like a V.)

The crucial feature of the above analysis that clinched the identification of the omega-minus is that it decays in three steps as it loses its three units of strangeness. Gell-Mann had been vindicated again. With the discovery of the omega-minus, Gell-Mann's noble Eightfold Way would now be the royal road to understanding the strong force.

Three Quarks for Muster Mark

> Three quarks for Muster Mark!
> Sure he hasn't got much of a bark
> And sure any he has it's all beside the mark.

James Joyce, *Finnegans Wake* (1939)[18]

The confirmation of Gell-Mann's remarkable predictions of new particles, first the eta (η^0) and then the omega-minus (Ω^-), was a convincing demonstration of the power of his ideas. He had discerned a symmetry in the properties of the particles that clearly played an important role in the physics of the strong force. But what was the origin of this symmetry? We know the answer to the equivalent question for the structure of the periodic table: the composition of atoms leads to the regularities in the properties of the chemical elements. The atomic number is equal to the number of protons in the nucleus of the atom, and this determines the number of electrons in the atom, which is ultimately responsible for the chemical properties of the atom.[19]

Gell-Mann now realised that the symmetry of the Eightfold Way could be understood in a similar fashion. The patterns of particles in his Eightfold Way scheme could be explained if all the strongly interacting particles (the proton, the neutron, the mesons and the others) had a composite

Answer to Puzzle 9

As described in the text, in the case of the octet of baryons the Eightfold Way symmetry is broken, and this produces a mass difference between the isospin multiplets when they are combined to form an Eightfold Way multiplet.

The difference in mass between the quartet of Δ resonances and the triplet of Σ^* resonances is

1,385 MeV – 1,236 MeV = 149 MeV

The difference in mass between the triplet of Σ^* resonances and the doublet of Ξ^* resonances is

1,530 MeV – 1,385 MeV = 145 MeV

which is almost exactly the same as the mass difference between the Δ and Σ^* resonances. Assuming that the mass difference between the Ξ^* doublet and Ω^- is similar, this gives a mass for the Ω^- of around

1,530 MeV + 147 MeV = 1,677 MeV

Based on the data available in 1962, Gell-Mann predicted a mass of 1,685 MeV for the omega-minus.[14] Modern measurements give the actual mass of the omega-minus particle as 1,672 MeV.[15]

structure. He was not alone in this observation. George Zweig, a researcher working at CERN in Geneva, independently came to the same conclusion.[20] Zweig called the subcomponents of the particles 'aces', but it is Gell-Mann's catchy terminology that has stuck. Gell-Mann was a great lover of exotic language, and was forever showing off his knowledge of obscure languages and dialects. He was just ten years old when his older brother Ben bought a copy of James Joyce's bizarre novel, *Finnegans Wake*.[21] Gell-Mann was fascinated by this hypnagogic excursion into a swirling maelstrom of language. A quarter of a century later, in 1964, he borrowed a word from the lines that open a chapter of *Finnegans Wake*, quoted at the head of this section, to name the ultimate constituents of matter. 'Quark' has since become almost a household word. Although Gell-Mann originally thought of quarks as useful mathematical tools that did not necessarily have any physical existence, it eventually became clear that they were a new class of fundamental particles. A deeper layer of reality had been revealed. Gell-Mann postulated the existence of three types of quark: the up quark, the down quark and the strange quark. These different types are now referred to as different flavours of quark.

Gell-Mann's idea was that mesons, such as pions and kaons, are formed from a quark and an antiquark bound together. For instance, the negatively charged pion is composed of an up quark and an anti-down quark. All particles that are formed from a pairing of a quark and an antiquark are today known as mesons. (The original definition of meson as a particle of intermediate weight is no longer applicable, as some mesons are far heavier than the proton.) We can see how the quark constituents of the mesons give rise to their properties from Figure 50, which shows the meson octet in terms of the quarks. The symbols for the various flavours of quark are as follows: the up quark is represented as 'u', the down quark is represented as 'd' and the strange quark is represented as 's'. A bar is placed over the letter to indicate an antiquark.

If particles are made of quarks, this immediately explains many of their properties. For instance, the strangeness charge is merely a measure of the number of strange quarks that a particle contains. Mesons with strangeness −1 contain one strange quark; mesons with strangeness +1 contain one strange antiquark. (It is simply a historical

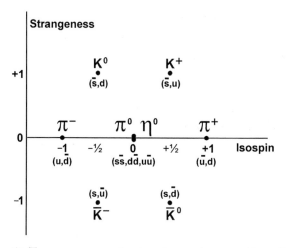

Figure 50 The meson octet, showing the quark composition of the mesons. The quark components of each mesons is indicated. For instance, in the top row the K-nought is formed of an antistrange quark and a down quark, whereas the K-plus is formed of an antistrange quark and an up quark. The π^0 and η^0 are composed of different combinations of up–antiup, down–antidown and strange–antistrange quarks.

accident that identifies the strange quark with a strangeness of −1 and its antiparticle with a strangeness of +1, rather than the other way round, which might seem more natural.)

Crucially, there is a second way in which particles are formed out of quarks in Gell-Mann's model. Three quarks may bind together to form a particle. These particles are the baryons. For instance, the proton and the neutron are baryons as they are each composed of three quarks. The proton is formed of two up quarks and one down quark (u, u, d); the neutron is composed of two down quarks and one up quark (u, d, d). The quark content of each particle in the baryon octet is shown in Figure 51; the quark content of each of the particles in the decimet is shown in Figure 52.

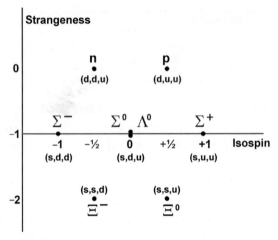

Figure 51 Baryon octet showing the quark composition of the baryons. The quark content of each baryon is shown in parentheses. For instance, the neutron, n, is composed of two down quarks and an up quark: (d, d, u).

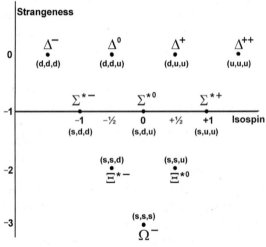

Figure 52 Decimet with quark content.[22] The quark content of each particle is shown in parentheses. For instance, the omega-minus is composed of three strange quarks: (s, s, s).

Puzzle 10

What would you expect the composition of an antiproton and an antineutron to be in Gell-Mann's quark model?

If we accept the quark hypothesis, the mystery of the Eightfold Way symmetry is resolved. For instance, isospin is just the symmetry under the interchange of the up and down quarks. If we swap all the up quarks with down quarks, then protons, which have quark content (u, u, d), are transformed into neutrons, with quark content (d, d, u), and conversely neutrons are transformed into protons. Isospin is a symmetry of nature simply because the up quark and the down quark have almost the same mass, and this explains why the proton and the neutron have almost equal mass.

The full Eightfold Way symmetry is an immediate consequence of forming particles by combining the three flavours of quarks and antiquarks. If all three quarks had the same mass, then the symmetry would be exact, and all the particles in an octet or the decimet would have the same mass. But the strange quark is actually a lot heavier than the other two quarks. It is this mass difference, of around

---

Puzzle 11

The proton is formed from two up quarks and one down quark, whereas the neutron is formed from two down quarks and one up quark: $p = (u, u, d)$; $n = (u, d, d)$. The proton has electric charge +1; the neutron has electric charge 0. These electric charges are the sum of the electric charges on the constituent quarks, so what are the electric charges of these two different flavours of quark?

Answer to Puzzle 10

The antiproton is composed of two anti-up quarks and an antidown quark. The antineutron is composed of two antidown quarks and an anti-up quark. Replacing each quark in the octet of baryons with its antiquark will produce the octet of the antiparticles of the baryons.

Answer to Puzzle 11

Look at the neutron first. If the sum of the electric charges on an up quark and two down quarks is 0, then the charge on the up quark must be twice that on a down quark, but opposite.

Now look at the proton. The down quark will cancel half the charge on one of the up quarks. Therefore the charge on the proton must be equal to one and a half times the charge on the up quark. The charge on the proton is +1, so the charge on an up quark must be $+\frac{2}{3}$.

We already know that the charge on the down quark is the opposite of half that on the up quark, so the electric charge of the down quark is $-\frac{1}{3}$. Incidentally, the fact that the omega-minus (Ω^-), whose electric charge is -1, is composed of three strange quarks, tells us immediately that the electric charge of the strange quark is also $-\frac{1}{3}$.

We are accustomed to the electric charge on the proton and the electron being the fundamental unit of electric charge, so it may come as a surprise that the charges we must assign to the quarks come in multiples of one-third of the charge on an electron.

150 MeV, that produces the mass differences between the isospin multiplets in an octet and in the decimet. Each additional strange quark increases the mass of a particle by

about 150 MeV, and this mass difference breaks the symmetry of the Eightfold Way.[23]

Viewed through the clarity of the quark model, the isospin and strangeness charges are revealed to be no more than accounting devices that keep track of the various flavours of quark. These charges are not really fundamental, they just reflect the types of quark from which the particles are formed. So from now on we can forget about isospin and strangeness and just think in terms of quarks. Conceptually, this represents a great simplification in our understanding of the physics.

If you were shocked to learn from the answer to Puzzle 11 that the quarks have fractional charges, then you are not alone. For many years there was a general reluctance among physicists to accept the existence of quarks as physical particles because of their fractional charges. Even Gell-Mann himself hedged his bets about whether quarks were real or just useful mathematical fictions. Like many great ideas, the quark model looks obvious in retrospect, but it took several years for the idea of quarks as real particles to become accepted. One of the defining features of a good physicist is an instinct to treat all new theories with a healthy degree of scepticism. If protons and neutrons were supposed to be made of quarks, then what physicists wanted to see was some sign of a free quark: surely, in the same way that electrons, protons and neutrons could be observed independently, it should be possible to find a quark existing independently, if only for a fleeting moment?

The Hunting of the Quark

The quark hypothesis provided such a neat explanation of the properties of the many new particles that were being discovered in accelerators that it seemed as though the simple rules of the strong-force game were about to be elucidated. The challenge to experimental physicists was to isolate the quarks and demonstrate that they really did exist. The unique characteristic of quarks is their fractional electric charge: protons, pions and all other known particles have electric charges that are multiples of the charge on an electron. But, as we have seen, the charge of an up quark is $+\frac{2}{3}$, the charge of a down quark is $-\frac{1}{3}$ and the charge of a strange quark is also $-\frac{1}{3}$. This property should make it relatively straight-forward to distinguish a free quark from any other particle. For instance, the width of a particle's track on a bubble-chamber photograph is determined by the particle's electric charge, so a fractionally charged particle should stand out, as it would leave a narrower trail.[24]

As soon as Gell-Mann had postulated the existence of quarks, the search began. Cosmic-ray photographs were re-examined for signs of quarks. Researchers around the world pored over photographs from particle accelerator experiments. Some began to search for fractionally charged particles in the places they supposed that quarks might accumulate if they were raining down in cosmic rays: in seawater, in oysters, in meteorites, in moonrock. But over forty years of searching has produced no evidence for the existence of free quarks. Yet high-energy scattering experiments in accelerators indicate very strongly that protons and neutrons really do contain point-like components with just the properties predicted by Gell-Mann's quark scheme. How can this apparent paradox be reconciled?

In particle accelerators, beams of very-high-energy particles are smashed into targets that contain protons. Some of the particles in the beam undergo head-on collisions with the protons. We might expect that, with sufficiently high-energy beams, we could knock a subcomponent, such as a quark, out of the proton. However, physicists have come to the conclusion that the force between quarks is so strong that it is impossible for a quark to escape imprisonment and exist as an independent free particle. They are effectively confined within protons and other particles. In the words of the Nobel laureate Sheldon Glashow, 'You can't even pull one out with a quarkscrew.'[25] This surprising property is called confinement.

We can get some insight into what is happening by comparing a meson, which is formed from a quark and an antiquark bound together by the strong force, with a hydrogen atom, which consists of a proton and an electron bound together by the electromagnetic force. Take the hydrogen atom first. We know that the electromagnetic force falls away sharply as the distance between two charged particles increases. At twice the distance, the electromagnetic force falls to a quarter of its original strength. If we imagine pulling the electron and the proton apart, we will have to put in energy to separate them, but once we have them far enough apart the electromagnetic force between them will have diminished to such an extent that we can consider the electron and the proton to be independent particles.

Now consider what happens if we try to separate the quark and the antiquark that form a meson. It takes a huge amount of energy to pull against the enormous force that holds the quark and the antiquark together. But the big difference in this case is that the force between the quark and the antiquark doesn't decrease as they are separated – it actually increases. This is a bit like pulling on a rubber band. The more we pull, the harder it is to stretch it further. As we pull our quark and antiquark farther apart, we must put in more and more energy to increase their separation. Very soon a point is reached where the energy we have put in is greater than the mass of a second quark–antiquark pair. At this point a new quark and a new antiquark may form. The original quark may now bind to the new antiquark, and the original antiquark may bind to the new quark. So instead of splitting the original meson into its quark and antiquark components, the energy that was expended has gone into generating two mesons out of one meson. In an accelerator experiment, the energy required to separate the quarks is introduced by firing a high-energy particle, such as an electron or a proton, at the meson. The energy would be transferred to the quark if it received a head-on impact from such a high-energy particle.

For this reason, it is believed that quarks are always found caged within composite particles such as mesons, protons or neutrons. Of course, this may appear to be rather an unsatisfactory explanation. If quarks exist, shouldn't it be possible to examine an individual isolated quark? There is an everyday example of forces that is so familiar that we take it for granted, but it operates in a similar way. A bar magnet has a north and a south magnetic pole. But we cannot isolate the north pole or the south pole. If we try cutting the bar magnet in half, we obtain two smaller bar magnets, both of which have a north pole and a south pole. It is as though we have created a new north pole and a new south pole out of thin air.

Although these arguments might make the claim of quark confinement seem less implausible, if we are to believe that this really is the way that quarks behave, then we need a theory that describes how the force between the quarks operates. It would be hard to fully accept the notion of quarks without such a theory.

Remarkably, physicists have indeed developed a theory that explains just how the force between quarks works, and it is ultimately the origin of the strong force that holds the nucleus of an atom together.

Colouring the Quarks

> Colour is sensibility in material form, matter in its primordial state.
>
> Yves Klein (1958)[26]

Although the quark model answered many questions, there were a few minor conundrums that it appeared to leave unexplained. A couple of these oddities provided clues that would lead to an understanding of the dynamics of quark interactions and ultimately explain how quarks are held together to form particles such as the proton. The strong force between protons and neutrons that holds an atomic nucleus together is a by-product of this force between quarks. It has a profound significance for the structure of matter.

By the 1970s, it was possible to generate sufficiently intense beams of positrons to conduct accelerator experiments in which electron beams were collided head-on with positron beams. Using the quark model, the proportion of the debris from such collisions that would be composed of quarks could be calculated.[27] When the experiments were done, there appeared to be about three times as much debris containing quark constituents as was expected from the calculations. It was as though there were actually three times as many types of different quark as had been thought to exist: three types of up quark, three types of down quark and three types of strange quark.[28]

The prediction of the omega-minus particle was one of the great successes of Gell-Mann's Eightfold Way scheme, but its quark composition also presented physicists with something of an enigma. Quarks are matter particles, and, like electrons, they must obey the exclusion principle. Just as two electrons cannot simultaneously exist in exactly the same state, neither can two quarks.[29] But the omega-minus particle is composed of three strange quarks (s, s, s), which all have their spins aligned in the same direction, so all three strange quarks are in the same spin state. But this is not allowed by the exclusion principle. This quirky fact about an obscure particle was threatening to undermine our whole understanding of quantum mechanics.

A number of theorists independently came to the conclusion that these oddities could be explained if quarks were carrying another new type of charge. Once again, it was Murray Gell-Mann who provided us with the memorable terminology that we now use. The idea is that quarks carry three additional charges, and these new charges are called red, blue and green. This simple manoeuvre removes the contradiction between the structure of the omega-minus and the exclusion principle. It means that the three strange quarks that form the omega-minus are not in the same state after all. One strange quark is carrying the red charge, the second strange quark is carrying the blue charge and the third strange quark is carrying the green charge. So, all three quarks are in a different state.

The extra property of colour also explained the apparently anomalous results of the high-energy scattering experiments. There was three times as much particle debris containing quarks as expected because the debris was formed from nine different types of quark rather than just three: red up quarks, blue up quarks and green up quarks; red down quarks, blue down quarks and green down quarks; and red strange quarks, blue strange quarks and green strange quarks.

Even so, the colour proposal seemed rather contrived. It did offer a solution to certain puzzles within the quark model, but at the expense of complicating matters in what appeared to be quite an arbitrary way. However, as we will see, this simple face-saving hypothesis turned out to be a big step towards understanding the interactions between quarks that is the essential origin of the strong force.

Yang–Mills Theory

Chen-Ning Yang was born in 1922 in Hefei, the capital of the Anhui province of China. He was the son of a maths professor and spent his early years on a university campus. His own university education took place during the Japanese occupation of eastern China, first at Beijing University and later at Kunming in Yunnan, when the university was evacuated from the occupied regions of the country. He went on to research various applications of group theory to physics, and in 1946 he left China to study for a PhD at the University of Chicago. He has remained in the United States ever since, becoming an American citizen in 1964. In the West, he is known as Frank Yang.

In 1954, Yang and the American theorist Robert Mills attempted to construct a theory of the strong force modelled on QED, the highly successful quantum theory of electromagnetism. QED has a symmetry group that is equivalent to the symmetry of a circle under rotations around its centre. It is this exact symmetry of QED that ensures that electric charge is conserved in all interactions. Yang and Mills realised that it might be possible to construct a theory of the strong force with a structure similar to that of QED. They formulated this theory as an elegant generalisation of Maxwell's equations. Instead of using the symmetry group of QED, they built their theory around the isospin symmetry of the strong force. Whereas the electromagnetic force arises from the exchange of photons, in their new theory the strong force was produced by the exchange of the three different types of pion. The theory had many nice mathematical properties. There were no arbitrary adjustable parameters, and the structure of the theory was completely determined by its symmetry group. It was like a perfectly cut, multi-faceted mathematical jewel. But unfortunately it did not fit the facts. One serious problem was that, according to the theory, the force-mediating particles had to be massless, like the photon, otherwise the diamond-like structure of the theory would contain a hidden flaw that would render it worthless. And while the mass of the pions is small, it is certainly not zero. So this theory could not be a serious contender for the correct theory of the strong force.

As we have seen, it turned out that the isospin symmetry of the strong force is explained simply by the fact that the particles that feel the strong force are built out of quarks. In particular, it derives from the similarity in the mass of the two lightest quarks, the up quark and the down quark. So ultimately, isospin symmetry is not responsible for the interactions between the quarks. However, the model built by Yang and Mills would not be forgotten, and eventually the importance of Yang–Mills type theories would become clear.

Almost twenty years later, in 1973, the German theorist Harald Fritzsch, his Swiss colleague Heinrich Leutwyler and Murray Gell-Mann together realised that with a suitable modification of the symmetry group, a much more promising

Yang–Mills type model could be constructed. Their work launched quantum chromodynamics, which is usually abbreviated to QCD, as a possible theory of the interaction between quarks. Inevitably, the name chromodynamics, derived from the Greek word *chroma*, meaning 'colour', was coined by Gell-Mann. The new theory was based on an exact symmetry under the interchange of the three colour charges carried by the quarks. Although colour symmetry was a completely different symmetry that was now proposed to explain the dynamics of the strong force, mathematically the symmetry of the three colours is described by exactly the same symmetry group as is Gell-Mann's Eightfold Way. The beauty of Yang–Mills theories is that they are very tight mathematical straitjackets. Once the symmetry group is determined, everything else follows inexorably – there is absolutely no room for manoeuvre. And that means that the theory is either right or wrong: it cannot be altered to fit the facts. This is exactly the sort of theory that physicists love. If it agrees with experiment, it is not because they have imposed their own preconceptions on the observations: it must be that the theory is revealing some fundamental truths about the way the universe operates.

So, how does the colour force described by QCD work?

The Psychedelic Force

The colour force binds three quarks together to form a proton or any other type of baryon. The use of the word 'colour' is by analogy with the mixing of red, green and blue light to form white light, which is how white light is produced on a television screen or computer monitor. It doesn't have anything to do with colour in the usual sense of the word. Colour in its everyday sense is, of course, a property of our perception of light, which has nothing to do with the strong force.

Quantum chromodynamics, or QCD, the theory of the colour force, has a structure similar to quantum electrodynamics, but with a number of important differences. Whereas interactions in QED depend on a single electric charge, in chromodynamics there are three different charges: red, blue and green. Combining three quarks, one with each of these colour charges, produces a particle, such as a proton, that is neutral with respect to the colour charges. In other words, the sum of a red charge, a blue charge and a green charge is no overall colour charge. A proton must contain one red quark, one blue quark and one green quark, so that as a whole the proton is colour-neutral.

Each of the red, blue and green charges also has a negative. These charges might well be called cyan, yellow and magenta if we chose to pursue the colour analogy further, but this would probably just cause confusion, so they are usually referred to as anti-red, anti-blue and anti-green. This gives us another way in which a colour-neutral particle might be constructed. A quark and an antiquark can be bound together, with the quark carrying one of the three colour charges and the antiquark carrying the corresponding anti-colour charge, so that the charges cancel and together there is no overall colour. For example, the quark might carry the red charge and the antiquark the anti-red charge. Alternatively, the quark might be carrying the anti-blue charge and the antiquark the blue charge. We know such particles, formed from a quark bound to an antiquark, as mesons. In this respect a meson is like the strong-force equivalent of a hydrogen atom, in which a positively charged proton is bound to a negatively charged electron. Overall, the hydrogen atom is electrically neutral.

The electromagnetic force is mediated by the exchange of photons between electrically charged particles, as we saw when we considered QED earlier. QCD operates in a similar way, with the colour force also being mediated by massless particles. These particles are known as gluons, because they are the particles that provide the glue that sticks the quarks together. Quarks are permanently immersed in a cloud of gluons that they are continually emitting and reabsorbing, but sometimes a gluon that has been emitted by one quark might be absorbed by another nearby quark. When this happens, some of the energy and momentum of the first quark is transferred to the second quark. It is this exchange of gluons that gives rise to the colour force between the quarks.

The techniques that Feynman invented for tackling QED can be extended for use in QCD. Figure 53 shows a QCD Feynman diagram representing an interaction between two quarks that scatter off each other. In the Feynman diagrams of QED, the solid lines with arrows on them represented electrons. Here, the solid lines with arrows on them represent quarks. The curly line represents a gluon that is being exchanged between the two quarks. (Compare this diagram with Figure 32 on page 178.)

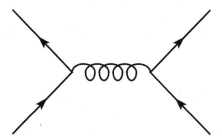

Figure 53 A QCD Feynman tree diagram.

From this simple diagram it might be assumed that all we need to do is replace electrons with quarks and photons with gluons, and we can complete any QCD calculation by converting it into the equivalent QED calculation. Unfortunately, things are not so straightforward. The difference in the structure of the symmetry groups of these two theories produces a fundamental difference between the electromagnetic and colour forces. Whereas there is only a single particle, the photon, that mediates the electromagnetic force, the colour force is mediated by eight types of gluon. The

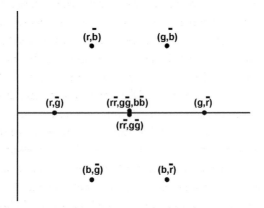

Figure 54 The gluon octet. Although it is a completely separate manifestation of the symmetry, the colour symmetry group is exactly the same as the symmetry group of the Eightfold Way. Its mathematics is therefore identical, and the gluons form an octet, just like the meson octet shown in Figure 50. For the mesons, the octet arises because there are three flavours of quark – up, down and strange – and each meson is formed from a quark–antiquark pair. The gluons form an identical octet because there are three colour charges – red, blue and green – and each gluon carries a colour charge and an anti-colour charge. The two gluons at the centre of the octet have no overall colour charge. They carry different combinations of red anti-red, green anti-green and blue anti-blue charges.

eight QCD gluons fit together into a colour-symmetry octet, as shown in Figure 54. This pattern should look familiar.

Each gluon carries two charges, both a colour charge and an anti-colour charge. The octet diagram shows the colour charge and the anti-colour charge of each of the eight gluons. For instance, the gluon represented by the point at the top left of the diagram carries a red charge and an anti-blue charge. This particular gluon will partici- pate in interactions as illustrated in Figure 55. On the left, the incoming red quark emits the red anti-blue gluon and is thereby transformed into a blue quark. This interaction conserves red charge because the red charge is transferred to the gluon. It also conserves blue charge, because the blue charge on the quark and the anti-blue charge on the gluon have been created simultaneously. Then, on the right, the red anti-blue gluon interacts with a blue quark. The anti- blue charge on the gluon cancels the blue charge on the quark and the red charge on the gluon is transferred to the quark. The overall effect of the exchange of the gluon is to transfer energy and momentum between the quarks and to swap over their colour charges. The same effect would be produced if the blue quark emitted a blue anti-red gluon

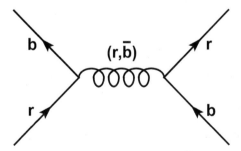

Figure 55 Gluon exchange between quarks.

that was absorbed by the red quark; the Feynman diagram does not distinguish between these two processes.

The gluon octet diagram looks familiar because the gluon octet has exactly the same structure as the meson octet of the Eightfold Way. The organisation of the mesons into an Eightfold Way octet is explained in the quark model by the fact that mesons are composed of a quark–antiquark pair, and there are three flavours of quark and antiquark: up, down and strange. The gluons form a similar octet because each type of gluon simultaneously carries one of the three colour charges and one of the three anti-colour charges. In both cases the symmetry group is the same. In the first case it is the symmetry under the shuffling of the three quark flavours; in the second it is the symmetry under the shuffling of the three colour charges. But whereas the flavour symmetry of the Eightfold Way is now known to be an approximate symmetry, arising simply from the near equality of the quark masses, the colour symmetry is believed to be a fundamental, exact symmetry of the universe.

It is this symmetry that is the key to the nature of the colour force. QED, the fantastically successful theory of electromagnetism, is built around a very simple group of symmetries that is equivalent to rotations around a circle. The theories that Yang and Mills modelled on QED were constructed using much bigger symmetry groups. The most important consequence of this is that the particles mediating the forces described by these theories behave in a very different way to the photon. Photons are electrically neutral, so they do not feel the electromagnetic force themselves and do not interact directly with other photons. Gluons, on the other hand, carry the colour charges, as indicated in Figure 54, so they feel the colour force themselves. This means that gluons interact with other gluons: a gluon can split into two gluons, or two gluons can combine to form a single gluon, and this greatly complicates the colour interactions and makes the colour force very different to electromagnetism.

Figure 56 shows several one-loop QCD diagrams. Again, the solid lines represent quarks and the curly lines represent gluons. The two diagrams on the left are familiar from our consideration of QED, and replacing quarks with electrons and gluons with photons would give equivalent QED diagrams. But the two QCD diagrams on the right have no QED equivalents. This is because the gluons carry the colour charges and so interact directly with other gluons. For

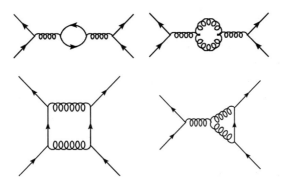

Figure 56 One-loop QCD Feynman diagrams.

instance, in the diagram at the top right, a quark emits a gluon, and this gluon dissociates into a pair of gluons which then recombine to form a single gluon, and this single gluon is absorbed by the second quark. As well as making QCD calculations even more formidable than QED calculations, the additional diagrams indicate that the force behaves in a completely different way.

It turns out that the colour force is actually quite feeble at very short distances, becoming strong only at longer distances. In this respect it is the complete opposite of the electromagnetic force. This is one of the great successes of QCD, because it is exactly what is observed in collider experiments. We can imagine the force between two quarks as the net result of a horrendously complicated exchange of multitudes of gluons that are simultaneously interacting with one another. This mass of tangled gluons effectively forms a tube of 'colour flux' that behaves a bit like a rubber band between the quarks, as illustrated schematically in the Figure 57. But our belief in the validity of QCD as an explanation of the strong force does not depend on such arguments. QCD is a Yang–Mills theory, so it has an extremely tight logical structure. The theory either stands or falls in its entirety. Ultimately, it must pass precise experimental tests or it will be discarded as a beautiful but false theory. So, what do the

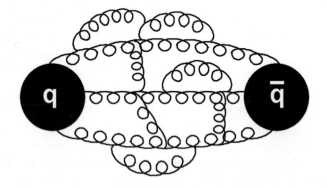

Figure 57 Schematic diagram of a complicated gluon exchange between a quark and an antiquark. The reason that QCD is so different from QED is that gluons, the particles that mediate the colour force, also feel the colour force themselves and therefore interact with other gluons.

results from the accelerator laboratories reveal about quarks and the glue that holds them together?

The Jet Set

Ever since Rutherford, physicists have sought ways to produce particle beams of higher and higher energy to enable them to probe the structure of matter at shorter and shorter distances. In the late 1960s and early 1970s, a series of experiments was performed at the Stanford Linear Accelerator (SLAC) in California with the aim of peering inside the proton to look for its hidden substructure. The method was very similar to Rutherford's alpha-particle experiments, though on an altogether grander scale. Proton targets were bombarded with beams of high-energy electrons, and the results showed that protons do indeed contain tiny hard components that were scattering the electrons. It is natural to infer that these nuggets within the proton really are Gell-Mann's quarks. So, even if it is not possible to extract and isolate individual quarks, we can still probe a proton and demonstrate that it contains these quirky constituents. The experiments conducted at SLAC and later accelerators suggest that typically about half the momentum of a proton is carried by their quark components, and the rest by the gluons that are flying around within the proton and holding the quarks together.

In 1978, a new accelerator called PETRA[30] was commissioned at the DESY laboratory near Hamburg in Germany. This facility was designed to study very-high-energy collisions between beams of electrons and beams of positrons. The beams circulate in opposite directions around the same ring, and at designated points they are focused and guided together. When a high-energy electron smashes head on into a high-energy positron, they mutually annihilate and their combined energy is released. The result, which was seen clearly for the first time at PETRA, is usually the production of two narrow jets of particles spraying out back to back from the impact point.

This is in agreement with what physicists expect from their understanding of QCD. Their interpretation is that a head-on collision of an electron and a positron results in their complete annihilation, and from the energy that is released a high-energy quark and a high-energy antiquark are produced. But, as we know, the quark and the antiquark carry colour charges, which means that the colour force operates between them. As they recede from each other, the energy contained in the colour interaction between them is converted into a shower of other quarks and antiquarks. All the quarks and antiquarks rapidly combine, so that their naked colour charges become hidden within colour-neutral particles, and these particles are the ones that are seen in the detector. The event appears in the detector as two narrow jets of particles emitted in opposite directions from the point of the electron–positron impact.

Occasionally, three jets of particles are produced. These events are even more interesting. PETRA produced the first such three-jet event in 1979. The QCD analysis suggests that occasionally either the quark or the antiquark produced in the electron–positron impact emits a gluon just at the moment that it comes into existence. The gluon results in a third jet of particles emanating from the point of the electron–positron impact.[31] By analysing the distribution of all the particles within a jet it is possible to distinguish between a jet that has formed from a quark and one that has formed from a gluon. It might not be possible to isolate a quark or a gluon, pick it up with tweezers and put it in a box, in the same way that an electron can be isolated and studied, but the production of individual quarks and gluons is effectively being seen in these jets.

As accelerator energies have increased, the tests of QCD have been refined further. Throughout the 1990s, the Large Electron–Positron Collider at CERN smashed beams of electrons head on into beams of positrons. In most of the collision events observed at LEP, either two or three complicated jets of particles were seen spewing from the collision zone. Figure 58 shows a reconstruction of one of these two-jet events.[32]

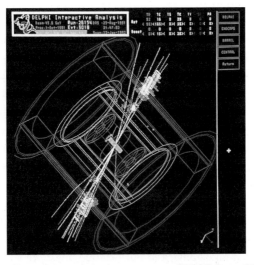

Figure 58 Reconstruction of a two-jet event at CERN's Large Electron–Positron Collider.[33] The collision of an electron and a positron within the DELPHI detector of LEP results in the creation of a quark and an antiquark, each of which immediately evolves into a jet of particles streaming from the impact point.

Experiments at the world's leading accelerators continue to confirm the predictions of QCD, but often it is just too difficult to perform the QCD calculations to work out exactly what they imply in a particular situation. QCD is much like the board game Go: its mathematical definition is very short and simple, but it results in extremely complicated interactions and it is incredibly difficult to work out their consequences accurately. However, the tremendous computer processing power now available to physicists has thrown them a lifeline and enabled them to model QCD on modern supercomputers. One of the aims of this number-crunching is to predict the masses of the particles formed out of quarks from first principles. This is similar to working out the energy levels in an atom – but vastly more complicated. When the computed masses of the various mesons and baryons are set against the values that are measured in particle accelerators, the agreement is very good[34]; typically, QCD predictions match experimental results to better than a 4 per cent accuracy.[35] Although this is not comparable to the incredible precision of the predictions of quantum electrodynamics, it is still very impressive. The accuracy of the calculations will increase as greater computing power becomes available and as the techniques for performing the calculations are further refined. Hopefully, the success of QCD will endure and the agreement with experiment will continue to improve. All the evidence suggests very strongly that quantum chromodynamics is the correct theory of quark interactions.

We can now go back at our original question: what is the origin of the force that binds protons and neutrons together in an atomic nucleus? Our modern understanding is that this force is a spin-off of the colour force between the quarks and gluons within the protons and neutrons. Although the resulting force – the strong force – is so strong that it over-whelms the electromagnetic repulsion between the protons and prevents an atomic nucleus from blasting itself apart, it is a mere shadow of the colour force. It is believed that a trace of the colour interaction leaks out of the protons and neutrons and manifests itself in a complicated exchange of mesons shuttling back and forth between the protons and neutrons. This meson exchange results in the attractive force that holds the nucleus together.

Gell-Mann and Complexity

In accordance with his life-long interest in diverse and com-plex phenomena, Gell-Mann founded the Sante Fe Institute in 1987. The Institute is devoted to the study of complex physical systems, with the aim of encouraging multidisci-plinary collaborations. The search continues there for the simple laws that are responsible for the complexity of the world that we see around us.

Gell-Mann was awarded the 1969 Nobel Prize in Physics 'for his contributions and discoveries concerning the classifi-cation of elementary particles and their interactions.'[36] In his acceptance speech, he explained the motivation of his quest to understand the forces of the universe:

> For me, the study of these laws is inseparable from a love of nature in all its manifestations. The beauty of the basic laws of natural science, as revealed in the study of particles and the cosmos, is allied for me to the litheness of a merganser diving in a pure Swedish lake, or the grace of a dolphin leaving shin-ing trails at night in the waters of the Gulf of California, or the loveliness of the ladies assembled at this banquet.[37]

This chapter has traced the path followed by physicists in their elucidation of the strong force. The complicated dynamics at the heart of matter are now understood in terms of the elegant theory known as quantum chromodynamics that accurately describes the colour force between quarks and gluons. QCD is a Yang–Mills theory built around colour symmetry. Remarkably, the other two forces that are important in particle physics – electromagnetism and the weak force – have now been united in a unified electroweak theory, and this theory is also a Yang–Mills theory. The next chapter tells the tale of this unification and the role of the Higgs particle in bringing it about.

Chapter Seven

THE MYSTERY OF THE SECRET SYMMETRY

> As theorists sometimes do, I fell in love with the idea
> [of spontaneous symmetry breaking]. But as often happens
> with love affairs, at first I was rather confused about
> its implications.

Steven Weinberg (1980)[1]

Picture Yourself on a Boat on the Grand Canal

Picture yourself on the Grand Canal, gliding along with your lover in your arms, absorbing the seductive charms of the Serene Republic. The waves lap against your gondola as the music of masked balls fades in the distance. As you pass the symmetrical facades of the palaces of the sensuous city of

Casanova, propelled by the rhythmic action of the gondolier, you gaze up at the starry heavens and ponder the meaning of the universe and the unity of the forces of nature. You might think that there is no connection between your gondola and the structure of these forces, but there is.

Almost all boats have a bilateral symmetry, which means that the left-hand side of the vessel is a mirror image of the right-hand side. Rowing boats have this bilateral symmetry, as do canoes, barges and motor boats. The same is true of most other types of vehicle: cars, railway locomotives, hovercrafts and aeroplanes. There is a very good reason for this. The propulsion systems that are used to move these vehicles forwards also exhibit this bilateral symmetry, which means that they propel the left-hand side of the vehicle in the same way that they propel the right-hand side. This explains why boats are built symmetrically. If the left side had a different shape to the right, but both sides were propelled forwards in the same way, the boat would not travel forwards in a straight line.

But the propulsion system of the gondola is unusual. This craft was the main method of transport through the canals of Venice for many centuries. The gondolier stands at the rear of the gondola and propels it by rowing with a single large oar that enters the water on the right-hand side of the boat. A casual glance might give the impression that the gondola is symmetrical. This raises the question of how the gondoliers can move their boats forward so effortlessly with their one-sided propulsion system. How are they able to glide along so elegantly, without jamming the canals of Venice with unruly gondolas continuously veering off course? A closer look at a gondola (see Figure 59) will explain the mystery. Over the years the traditional design of gondolas has evolved to enable the gondoliers to propel them efficiently. In order

Figure 59 Photograph of a gondola taken from the front of the boat. The curvature of the hull is quite obvious.

for the gondola to be perfectly balanced, and so that it is propelled forwards without any deviation to the left-hand side, the design of the gondola is markedly asymmetrical.

Puzzle 12

Apart from the gondola, what other small boats do not have bilateral symmetry, and why?

Answer to Puzzle 12

Yachts and other sailing boats are propelled by the wind. The direction in which the wind is blowing is typically not the same as the direction in which the sailor wishes to travel. With a symmetrical boat, with square sails set perpendicular to the long axis of the hull, it would be possible to travel only in the same direction as the wind. However, altering the angle of the sails with respect to the rest of the boat enables movement in other directions. In this way the bilateral symmetry of the boat is broken. With yachts and other small boats, the symmetry is usually also broken by the shape of the sails themselves.

The hull is not straight but curves towards the right, so that an unladen gondola will list to this side. This counterbalances the push of the single oar which would otherwise tend to direct the boat to the left. In short, gondolas are constructed asymmetrically to compensate for the asymmetry in their method of propulsion.

The romantic boater on the River Cam in Cambridge resembles the gondolier as he propels the punt with a pole that enters the water on one side of the boat. The punt has yet to evolve an asymmetry in its design to compensate for this. However, the experienced punter, maybe a research physicist from the Cavendish Laboratory enjoying a leisurely after-noon in Grantchester Meadows, is able to propel the punt in a straight line by combining two motions into a single stroke – a push off the bottom of the river and a sideways sweep of the pole through the water behind the boat, levering the pole on the hip. The asymmetry of the propulsion system becomes obvious in midsummer when tourists take out a punt for the first time, and the river is blocked by a chaotic logjam of punts zigzagging along the river.

The gondola is an example of a broken symmetry. It is almost symmetrical; and for many purposes it is simplest to imagine that it is symmetrical. Canaletto, when painting a majestic view along the Grand Canal, probably would not have worried about the lopsided shape of the gondolas and would have depicted them as though they were as symmetrical as any other boat. Gondolas are symmetrical to a good approximation. Nevertheless, their asymmetry is certainly important to the gondolier as it affects the way that his craft travels through the water.

Abel Was I Ere I Saw Elba

Many effects can be described in terms of broken symmetries. The application of forces can often be considered in this way. If forces are applied in a perfectly symmetrical way, all the forces will balance and there will be no overall force. For example, imagine a volatile mixture of gases exploding inside a perfectly spherical metal chamber (see Figure 60). Assuming that the metal sphere is strong enough to withstand

Figure 60 Symmetry breaking in a combustion chamber. Left: The symmetry is unbroken and there is no overall force acting on the sphere. Right: The symmetry is broken; now, not all the forces are balanced, so the sphere will accelerate upwards.

the blast, the motion of the sphere will not change, because the forces imparted by the expanding gases within the sphere will be spherically symmetrical. For every region of the sphere, the force of the expanding gases will exactly balance the force of expanding gases on the diametrically opposite region of the sphere, so all the forces will balance and there will be no overall force applied to the sphere. (The result will be that the metal sphere heats up and eventually this heat will dissipate.)

Now imagine what would happen if we made a hole on one side of the metal sphere before detonating the explosion, as shown on the right of Figure 60. The hole will break the spherical symmetry, and so we should expect that the forces will no longer exactly balance. And this is indeed the case. When the gases explode there will now be a region opposite to the hole where the outward force on the sphere will not be matched by an oppositely directed force. There is no force in the opposite direction, because the gas that is expanding in this direction will escape through the hole in the sphere and will therefore not impart a force on this region of the sphere. This unbalanced force will propel the sphere in the opposite direction to the gas that is escaping through the hole. What we have reconstructed in this mental image is the combustion chamber of a rocket engine: by breaking the spherical symmetry, an unbalanced force is produced that can propel a rocket.[2]

Can a Symmetrical Cause Have an Asymmetrical Effect?

The fundamental laws of the universe are, by definition, complete and self-contained. Throughout the twentieth century, physicists found evidence for symmetry in every corner of physics. The architecture of the universe really is built on perfect symmetries, and any fundamental theory of the forces of

---

Puzzle 13

What are the symmetries of each capital letter in the alphabet? Which capital letters are symmetrical when reflected in a vertical mirror down their centre? Which capital letters are symmetrical when reflected in a horizontal mirror across their centre? Which letters are symmetrical when rotated by 180 degrees or half a complete rotation around their centre? Which letters have no symmetry (in a sans-serif font)?

---

nature must incorporate these symmetries. But this presents theorists with a serious dilemma. If the ultimate laws of the universe derive from exact symmetries, then these symmetries cannot be flawed or broken. For the universe to be governed by laws built on beautiful symmetries that are almost but not quite true seems philosophically implausible and unsatisfactory. On the other hand, most traditional reasoning has suggested that exact symmetry can only lead to perfect homogeneity and a totally uninteresting and sterile universe where nothing happens. Somehow, some of the symmetry must disappear.

The gondola and the rocket engine illustrate the effects of symmetry breaking, but in each case the symmetry was broken by human intervention. When considering the ultimate laws of the universe as a whole, however, and the fundamental symmetries they encapsulate, there is no possibility of any outside influence breaking the symmetries because the universe is all there is. If symmetry is to be broken, for example the symmetry between the electromagnetic and weak forces, it must happen without any outside interference. Therefore, if theorists are to understand the mysteries of the universe in terms of symmetry breaking, they must find a way in which nature can be described by laws that are perfectly symmetrical

---

Answer to Puzzle 13

The letters that are symmetrical when reflected in a vertical mirror are: A H I M O T U V W X Y.

The letters that are symmetrical when reflected in a horizontal mirror are: B C D E H I K O X. (The exact symmetry of B and K require that the two lobes of the B are the same size and, in the case of K, that the two arms meet at the centre of the upstroke.)

Each letter that is symmetrical in both vertical and horizontal mirrors is necessarily also symmetrical when rotated by 180 degrees around its centre. These letters are: H I O X.

Three letters are not symmetrical under any reflections, but are symmetrical when rotated by 180 degrees around their centre. These letters are: N S Z.

The other five letters are not symmetrical under any reflections or rotations: Q F G J P.

---

and yet lead to effects in which some of the symmetry is lost. In previous ages, this would have seemed a great paradox.

This can be illustrated by a famous argument known as Buridan's ass, devised by the fourteenth-century French logician Jean Buridan. Buridan theorised about mechanics, in particular the behaviour of projectiles and the concept of impetus, which was an early attempt to define momentum. Buridan's discussion of the ass has not survived in any of his writings, so we do not know the true context of his argument. However, 'Buridan's ass' is usually thought of as a caricature of an indecisive person. The ass is standing in a symmetrical position, precisely halfway between two exactly equivalent bundles of equally delicious hay, one bundle to the left and one to the right. According to Buridan, since the position of the ass is exactly symmetrical with respect to the two bundles of hay, it cannot make a rational decision to move in either direction, and so dies of starvation in the middle. Buridan seems to be following a line of reasoning that can be traced back to some of the ancient Greek philosophers and can be summarised thus: if there is no sufficient reason for one thing to happen instead of another, then neither will happen. This sounds reasonable enough when presented as a philosophical argument, but is it genuinely how things work in the real world?

We will probably never know whether Buridan really was discussing the paradoxes of human indecision, or whether he had in mind a physical situation in which forces are balanced in a symmetrical way. However, we can consider a symmetrical set-up which is a mechanical version of his argument. A metal needle, balanced vertically above its pivot and free to rotate about the pivot, is placed precisely halfway between two equally powerful magnets, as shown at the top of Figure 61. According to Buridan's reasoning, the needle should remain where it

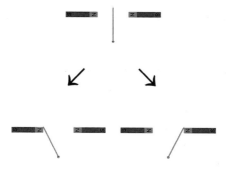

Figure 61 Above: A pivoted metal needle, viewed from above, is positioned between the north poles of two magnets. Below: This configuration is symmetrical, but unstable. The needle cannot remain balanced between the two magnets, but will rotate to the left or to the right. Both outcomes are equally likely, and either will break the symmetry.

is because the pull on it to the left exactly equals the pull on it to the right. However, this ignores the fact that the arrangement is unstable. The tiniest fluctuation within the needle or the magnets, or in the air between them, will tip the balance, and once the force towards one of the magnets is slightly larger than the force towards the other, the symmetry is broken and the needle will rotate either to the left or to the right. The needle is rather like a pencil balanced on its tip. We can imagine balancing a perfectly symmetrical pencil on its tip so that there is no reason why it should fall in one direction rather than another, but we know that in reality the pencil in this state is unstable and it will fall. As soon as it begins to fall, the symmetry is broken and it continues to fall in the same direction.

It is clear that, in the real world, the general conclusion that a symmetrical cause will always lead to a symmetrical effect does not necessarily hold. So, in principle at least, it is possible that the laws of the universe are perfectly symmetrical, even though this symmetry is not always apparent when we look at natural phenomena.

Skating on Thin Ice

A Dutch ambassador, who entertaining the king of
Siam with the particularities of Holland, which he was
inquisitive after, amongst other things told him that the
water in his country would sometimes, in cold weather,
be so hard that men walked upon it, and that it would bear
an elephant, if he were there. To which the king replied,
Hitherto I have believed the strange things you have told
me, because I look upon you as a sober fair man, but now
I am sure you lie.

John Locke, *An Essay Concerning Human Understanding* (1690)[3]

Most of us are very familiar with frosty winter scenes where snow and ice cover the ground, ponds and lakes are frozen, and dripping icicles hang from the trees, so the response of the King of Siam to the tale of the Dutch ambassador may seem rather amusing. It is easy to take for granted the familiar but remarkable transformation of a liquid into a solid as it freezes. But imagine seeing this happen for the first time, and the reaction of the king seems rather less absurd. Suddenly, as the temperature of the liquid water falls, a critical temperature is reached and the water rapidly transforms into a rigid solid, its molecules spontaneously organising themselves and locking into place to form an ice crystal.

The miraculous transformation that turns liquid water into ice is known to physicists as a phase transition. Many other phase transitions have been identified in which, at a well-defined temperature, the structure of a material changes completely. They include the condensation of a gas into a liquid, and the transformation of liquid helium into a superfluid. There is also the transition of a normal conductor into a superconductor, which happens in a range of materials that lose all electrical resistance at low temperatures. There are many others, and we will encounter some of them in this chapter.

On the face of it, these transformations all appear very different, but in the 1930s the Russian theoretical physicist Lev Landau realised that they all have one thing in common. In each case, the transformation of the material is accompanied by a loss of symmetry, and this insight was the first step towards building a mathematical model of phase transitions.

Landau's theory represents one of the most important insights in the whole of twentieth-century physics. It is remarkable because, as we will see, his reasoning was very general and independent of the details of any particular physical system. For this reason, it applies to many different areas of physics. There can be few occasions where pure reason has led to so great a pay-off.

The Theoretical Minimum

Lev Landau lived in a small Moscow apartment on two floors. The walls of the room on the upper floor were lined with bookcases, there was a desk by the window, and in the middle of the room was a sofa where the reclining Landau would conduct most of his research into theoretical physics. His compatriot George Gamow used to joke with his students that physics was a great career because you could settle into a comfortable armchair, lean back and close your eyes … and no one would know whether you were working or not.

Landau's formative years as a physicist were spent with Niels Bohr at his institute in Copenhagen. Landau believed that theorists should absorb mathematical techniques and fundamental physics to the point where they became second nature, as only then could they free their conscious faculties to explore physics creatively. From his early years as a researcher he began to develop an examination to assess the potential of his younger colleagues. There was no course, so the candidates had to rely on self-study. The examination became known as the Landau Theoretical Minimum, as it implied that any theorists who could not hurdle this barrier lacked the potential to become a top researcher. Candidates sat at the desk in Landau's apartment. The exam consisted of nine papers, some of which took seven hours to complete. In twenty-five years there were just forty-one theorists who passed.[4] The reward for successful candidates was that they earned the attention of Landau. The value of this prize can be judged by the fact that this elite group became the leading researchers of the Soviet Union, with several winning Nobel Prizes in Physics. The name of the exam – the theoretical minimum – was a punning reference to Landau's theory of phase transitions, as will soon become clear.

A Magnetic Attraction

Landau's theory of phase transitions is easiest to understand in its application to magnets. Most of us first encounter magnets and magnetic compasses at an early age. Einstein's first experience of magnetism came when he was no more than four or five years old:

> He was sick in bed one day, and his father brought him a compass. He later recalled being so excited as he examined its mysterious powers that he trembled and grew cold. The fact that the magnetic needle behaved as if influenced by some hidden force field, rather than through the more familiar mechanical method involving touch or contact, produced a sense of wonder that motivated him throughout his life.[5]

This permanent magnetism, as it is called to distinguish it from the magnetism produced by an electric current, is a property of iron and a few other metallic elements. As iron is far and away the most common of these elements, scientists know the effect as ferromagnetism, which is derived from *ferrum*, the Latin word for iron. Other ferromagnetic elements are nickel, cobalt and gadolinium. The world's strongest permanent magnets are made of an alloy of neodymium, iron and boron, known as NIB. According to the chemist John Emsley,

> NIB magnets are so powerful that those handling them must wear protective glasses – they fly together with such force that they can shatter and send splinters flying in all directions. At times young people have used these industrial magnets to attach ornaments to their cheeks by putting one of the small magnets on the inside of the mouth. However, the magnet and ornament have then proved impossible to pull apart, sometimes necessitating a visit to a hospital for surgical removal.[6]

Each of these ferromagnetic materials undergoes a phase transition at a temperature known as its Curie temperature, after the French scientist Pierre Curie, the husband of Marie Curie. For instance, the Curie temperature of iron is 1,043 K. Above this temperature a lump of iron will lose its magnetism, while below this temperature the iron will spontaneously magnetise. The Curie temperature of the silvery metal gadolinium is 20 °C (293 K), which is room temperature.[7] This is very convenient for a first-hand experience of the transformation in its physical properties. A piece of gadolinium cooled in an ice bucket will spontaneously magnetise. If it is then held in the hand until it returns to room temperature, its magnetism will suddenly disappear.

Spontaneous Symmetry Breaking

Landau had a remarkable intuition for how complicated physical systems behave. At times, his insight appeared almost uncanny. Such was his belief in his methods that he would pursue an argument through many ingenious steps until he reached a final, simple conclusion.

We can illustrate the application of Landau's ideas to ferromagnetism with a simple toy model. Imagine a collection of atoms arranged regularly in a plane (see Figure 62). Each atom has a small magnetic field that is constrained to point in one of two directions, either upwards or downwards. Both directions are completely equivalent, but the combined energy of two neighbouring atoms is slightly lower if their magnetic fields are pointing in the same direction. The magnetic fields can be thought of as 'sticky', as there is a force between them that makes them tend to align. At high temperatures, thermal vibrations vigorously jostle the atoms around, and their magnetic fields are continually being flipped. So, although there might be a tendency for neighbouring atoms to align their magnetic fields, any such alignments are rapidly disrupted. In this state the material will not exhibit a magnetic field macroscopically, as the randomly oriented magnetic fields of all the atoms cancel one another out. When the temperature is lowered, a point will eventually be reached where the thermal vibrations are no longer strong enough to disrupt the alignments, and suddenly all the magnetic fields spontaneously align in the same direction. The material will now have a macroscopic magnetic field. It has become a magnet.

The next step is to understand this behaviour in terms of symmetry. In everyday language, the words 'symmetry' and 'order' are considered almost synonymous – which is unfortunate, because when physicists use these terms their meaning is virtually opposite. This is not physicists being deliberately obtuse. When thought about in the right way it makes perfect sense, as a simple analogy should make clear. Take a class of ten pupils and number them 1 to 10. There

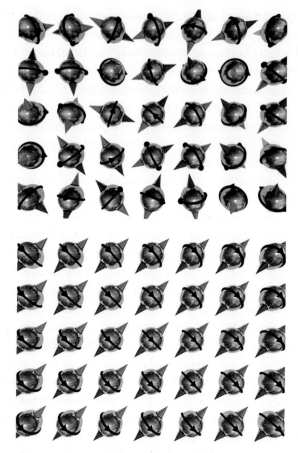

Figure 62 A lattice of iron atoms above and below the Curie temperature. The spheres represent the atoms, and the cones indicate the direction of the atoms' magnetic fields. Above: A snapshot of the lattice above the Curie temperature. The magnetic fields are constantly being rearranged, and are randomly oriented by the continual thermal buffeting in the lattice. There is no overall magnetic field. Below: When the lattice falls below the Curie temperature, all the individual magnetic fields become aligned in the same randomly selected direction. The alignment of the magnetic fields on the atomic scale produces a macroscopic magnetic field.

are many ways in which this can be done. Allow the pupils to keep randomly swapping numbers. This represents a symmetrical state. Over a period of time, each pupil has an equal chance of holding each number. Now imagine taking ten pupils and giving each pupil the same number, say 7. This represents an ordered state, as one of the ten numbers is now special. It is ordered because all the pupils have the same number. In general, high symmetry implies low order, and high order implies low symmetry.

How should our toy model of a magnet be interpreted in terms of symmetry? Above the transition temperature, the direction of the magnetic fields of the atoms are arranged completely at random. The magnetic field of each atom is equally likely to be pointing in either direction, so the system does not distinguish between the two directions, as they are completely equivalent. In this state the system is symmetrical with respect to the up and down directions, but below the transition temperature this symmetry has disappeared as the magnetic fields of all the atoms are now pointing in the same direction. The loss of symmetry below the phase transition is equivalent to a gain in order, as illustrated in this toy model. Above the transition temperature, the fields are oriented at random and they are completely disordered – this is symmetrical because there is an equal chance that each field will be pointing in either direction; below the transition temperature, the fields are aligned and form an orderly array – this is asymmetrical because one direction is picked out as being special.

This was the key to Landau's analysis. He suggested that there must be a parameter that measures the order. Above the transition temperature this parameter must be zero – as the system is completely disordered. Below the transition temperature the order parameter must take some non-zero value, a measure of the order that has spontaneously arisen. In the case of a magnet, the order parameter is called the magnetisation. (The magnetisation is equal to the proportion of the magnetic fields that are pointing upwards minus the proportion that are pointing downwards.) Above the transition temperature the magnetisation is zero, as there are equal numbers of the magnetic fields pointing in both directions; below the transition temperature the magnetisation is non-zero, because the magnetic fields have become aligned, so there are more pointing in one direction than in the other.

The total energy of the system depends on the arrangement of the magnetic fields, so the next step in Landau's analysis was to deduce an expression relating the energy to the magnetisation. This was very important because, as we have seen, physical systems always tend to fall into the lowest energy state available to them, so Landau's expression would determine the state the system would find itself in. His crucial insight was that the expression for the energy must respect the symmetry of the system. In our toy model we have assumed that there is perfect symmetry with respect to the up and down directions: if we flip the direction of all the magnetic fields, the energy must remain exactly the same. We can plot a graph, as in Figure 63, of how the energy changes as the magnetisation changes. The symmetry of our system implies that the graph must be unchanged if we change the magnetisation from M to $-M$. In other words it must be symmetrical when reflected in the energy axis. This dramatically reduces the number of possible shapes for the graph (for more details, see the Appendix). Indeed, Landau reasoned that there could be just two. The first possibility was that it might take a simple U shape, as shown in Figure 63. It is clear from this graph that the energy takes its lowest possible value at the bottom of the U, the point where the magnetisation is zero. The only other possible shape of the energy graph is shown in Figure 64. There are now two points at which the energy attains its minimum value. At one of these

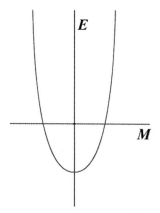

Figure 63 A graph of the energy *E* plotted against the magnetisation *M*, where the energy takes its lowest value for zero magnetisation.

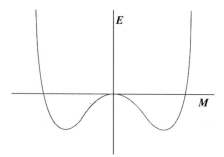

Figure 64 A graph of the energy *E* plotted against the magnetisation *M* where the energy now takes its lowest value at two different points, at one of which the magnetisation takes a positive value, and in the other a negative value.

points the magnetisation takes a positive value, and at the other a negative value.

Landau could now interpret the graphs in terms of the physical state of the magnet. He reasoned that the first graph must represent the energy just above the transition temperature, and the second graph must represent the energy just below the transition temperature. Landau's description of the phase transition was that above the transition temperature the system is in its lowest energy state, which is the lowest energy point of the curve in Figure 63. At this point the magnetisation is zero, which indicates a state in which the magnetic fields are oriented at random with respect to the up–down axis, so that there are equal numbers pointing upwards and downwards. However, as the temperature falls, the shape of the graph changes, and below the transition temperature there are two points of lowest energy, as in the curve in Figure 64, and the system must fall into one of these states. They are both equivalent, so the choice is made completely arbitrarily by random fluctuations within the system as the temperature falls below the transition temperature. In one minimum energy state all the atomic magnetic fields will be pointing upwards; in the other they will all be pointing downwards. In either of these states the system is now ordered, with all the magnetic fields aligned, and the up–down symmetry is lost – or at least hidden.

Landau's expression relating the energy to the order parameter is applicable to a whole range of phase transitions, and the energy graphs generalise in a natural way to higher-dimensional systems. For instance, if the magnetic fields were free to point in any direction in a plane, rather than being constrained to lie along a single axis, then below the transition temperature the energy graph would look like Figure 65. This is known as a Mexican hat diagram. The lowest energy states now form a circle within the brim of the hat, and at each point on this circle the energy of the magnetic system is as low as possible.

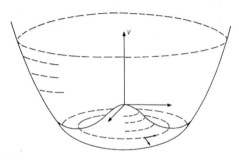

Figure 65 The Mexican hat diagram – the equivalent of Figure 64 for a two-dimensional system.

The spontaneous magnetisation of a ferromagnet, such as iron, that we have been considering in our simple model, illustrates a concept that is very important in modern physics – spontaneous symmetry breaking. Although the universe may be constructed in a very symmetrical way, allowing us to describe it with elegant symmetrical laws, sometimes this symmetry manifests itself only at high temperatures. As the temperature falls, matter will spontaneously reorganise itself into a more stable lower energy state in which some of the symmetry is lost. In fact, the symmetry described by the laws is really still there, but it is hidden. This is like our DNA, which codes for the bilateral symmetry of our bodies. We might reduce this symmetry by wearing an earring in one ear or by tattooing one arm and not the other. The traumas of life may even leave us with a scar on one cheek that breaks the natural bilateral symmetry of our face. But we retain the genes within all our cells that form the instruction set for developing a symmetrical body, and we will pass these genes on to our offspring. Similarly, the symmetry within the ferromagnet continues to play an important physical role, but it is not apparent at low temperatures. Although physicists usually refer to this effect as spontaneous symmetry breaking, many prefer to describe it as 'hidden symmetry'. At very low temperatures, matter behaves in surprising and unfamiliar ways that can often be understood in terms of spontaneous symmetry breaking.

Supercool

The science of low temperatures is known as cryogenics. This name is derived from the Greek words *kryos* and *genesis* and literally means 'to produce freezing cold temperatures'. The word *kryos*, which means 'frost' or 'icy cold', has also evolved into the modern English word 'crystal', as in 'ice crystal'. At atmospheric pressure, the boiling point of helium is 4.2 K, or just 4.2 degrees above absolute zero, which is lower than the boiling point of any other substance. In 1904 the Dutch physicist Heike Kamerlingh Onnes founded a cryogenics laboratory in Leiden dedicated to exploring the new frontier opened up by low-temperature physics. This was the first research laboratory to be organised on an industrial scale, and it would soon become the world leader in this branch of physics. Kamerlingh Onnes thought long-term, and even established schools of glass-blowers to provide equipment for the laboratory. In the early years of the twentieth century, a number of laboratories were engaged in a race to liquefy helium. The Leiden laboratory was the eventual winner, in 1908. But what no one realised until much later was that if liquid helium is cooled further, it undergoes a much more dramatic transformation.

Pyotr Kapitsa arrived in Cambridge as an émigré from Soviet Russia in 1921 and persuaded Rutherford to take him on as a research student at the Cavendish Laboratory. Initially Kapitsa was to stay for just one year, but as time went on he always managed to extend his stay. By 1929 he had come to an agreement with the Russian authorities that he could remain in Cambridge indefinitely on the understanding that he would make an annual visit back to his homeland, where he would act as a scientific advisor.

Kapitsa proved to be extremely adept at applying state-of-the-art engineering to tackle the questions of fundamental physics. He devised equipment to generate intense magnetic fields, and constructed innovative equipment for low-temperature research. His ingenious design for a pump to liquefy helium is still the basis for all modern high-capacity helium liquefiers today. By the early 1930s, with the backing of Rutherford, Kapitsa had established the Mond research laboratory, housed in a new building in the courtyard of the Cavendish Laboratory. He commissioned the sculptor Eric Gill to carve a crocodile into the brickwork of the building, where it can still be seen today. Kapitsa's mischievous explanation was that it was simply a representation of Rutherford. Kapitsa was generally considered to be in line to succeed Rutherford as head of the Cavendish. He continued to make his annual trip to Russia, but in 1934 while he was in Moscow he was shocked to discover that his passport was to be confiscated and he would not be allowed to leave the country. Rutherford and Dirac interceded on his behalf, but to no avail. Kapitsa would not see Cambridge again until 1966. For two years he refused to continue his physics research and was held virtually under house arrest.

From this time onwards, Kapitsa wrote pointed letters to Stalin. He never received a reply, but he knew that the letters were reaching their intended recipient as from time to time he could perceive certain changes that could only be in response to what he had written.[8] Sending these outspoken letters was brave, even rash, but Stalin seems to have treated Kapitsa with a certain amount of respect and even offered him some protection from the worst excesses of Beria, the psychopathic head of the Secret Police, the NKVD. By 1937, Kapitsa had grown to accept his changed circumstances. He had even been granted his own institute in Moscow, and much of his equipment had been transferred to him from Cambridge. One of the physicists who joined Kapitsa at the new institute was the young theorist Landau.

Kapitsa's return to physics paid immediate dividends. In 1937 he discovered that when liquid helium is cooled below 2.17 K it undergoes a remarkable transformation: suddenly, all resistance to its flow disappears. A container that holds normal liquid helium perfectly well will suddenly spring numerous leaks when it is cooled below the transition temperature, as liquid helium seeps out through ultramicroscopic pores in the container walls. Kapitsa named this phenomenon superfluidity. Superfluid liquid helium exhibits many strange and wonderful properties. For instance, a beaker containing superfluid helium will empty itself with the liquid flowing up and over the sides of the beaker. Placing a thin glass tube in superfluid helium has an even more dramatic result: the liquid spouts out of the top of the tube, and this is known as the fountain effect.

Another consequence of the loss of resistance is that super-fluid helium is an excellent conductor of heat. This makes the change of phase to superfluid helium obvious to see. Normal liquid helium bubbles vigorously, just as water does close to *its* boiling point. The bubbling is caused by variations in temperature within the liquid, as helium gas forms in regions of the liquid that are warmer than average. As the helium is cooled below the transition temperature it suddenly becomes completely still. The heat conduction in the superfluid is so good that any temperature variations immediately dissipate without any bubbles forming.

The year 1937 was the height of the Great Purge, in which Stalin's secret police arrested many of the leading citizens of the Soviet Union. Landau had not been at the Kapitsa Institute for long when he too was arrested. Kapitsa wrote to Molotov the same day in an attempt to restrain the authorities, requesting that they be lenient with the twenty-nine-year-old theorist and excuse his youthful exuberance.

One year later, Landau was still in prison, and his incarceration had taken its toll. Kapitsa continued to write letters in an attempt to secure his release. Finally, on 6 April 1939, Kapitsa wrote to Stalin's protégé Vyacheslav Molotov, explaining that he had made remarkable discoveries in his low-temperature research and his study of liquid helium, but that he needed some theoretical assistance: 'In the Soviet Union, it is Landau who has the most expertise in this domain; unfortunately, he has been in prison for an entire year.'[9] He then insisted that Landau could not possibly be a threat to the Soviet state, because he would be too engrossed in his theoretical work. This approach finally proved effective. Kapitsa gave Beria a personal guarantee for Landau's future conduct, and the half-dead Landau was released.

Landau had published a number of papers relating to his theory of phase transitions in the year before his imprisonment, but he appears to have developed his ideas much further during his captivity, as the most important applications of the theory date to the period just after his release. Incredibly, on regaining his freedom Landau immediately answered Kapitsa's prayers by producing a successful theory of superfluidity. Landau would be rewarded with the 1962 Nobel Prize in Physics for his work on superfluid helium; Kapitsa won the same prize sixteen years later for his own low-temperature research. Landau's next target would be another weird and wonderful low-temperature phenomenon. Over the following turbulent decade, his ideas would be developed in collaboration with his colleague Vitali Ginzburg.

Resistance is Futile

Many solid materials form crystal lattices in which some of the loosely bound outer electrons of the atoms escape and form a sea of electrons that are free to roam throughout the lattice. These materials are metals. The free electrons are responsible for the familiar properties of metals. Metals are very good conductors of electricity because the free electrons can flow through the metal. They also conduct heat well, as heat is also transported by the flow of electrons. This is, of course, why we use copper wires to make electric cables and why pans are made from aluminium or stainless steel.[10] However, the flow of electrons through a metal is not completely unimpeded. The atoms are continually jostling about, and these vibrations can give a passing electron a kick which alters its course through the material. The electrons may also interact with other electrons, and this again affects their flow. Both the scattering by the vibrations of the atoms and the scattering by other electrons hinders the flow of electricity and this is the microscopic origin of electrical resistance. Each time an electron is scattered it will lose some of its energy, which will be released as heat, thus raising the temperature of the electrical wire or whatever device the electricity is running through. It is because of this loss of heat energy that energy has to be continually added to maintain the flow of the electric current. If there were a way to enable the electrons to flow without undergoing any scattering, there would be no resistance and therefore no energy loss, and the current would run for ever. All electrical devices generate heat, and that can be a serious problem for devices as small as a computer chip. PCs have to contain fans to keep the processors from overheating. On the other hand, the heat is clearly beneficial if the electricity is powering a cooker or an electric heater. But this heat generation can be quite a headache when we receive our electricity bill. Overall, about 8.5 per cent of the electrical energy that is generated in power stations in the UK is lost as heat during its transmission over the National Grid, and that is equivalent to the total output of a couple of large power stations.[11] If the National Grid could transmit electricity without any loss through heating, a couple of Britain's most polluting power stations could be closed – which would reduce the cost of electricity and significantly lower the UK's carbon emissions.

An electric current that runs for ever without any energy loss sounds like science fiction – but it isn't, as Kamerlingh Onnes discovered. As temperature decreases, the atoms in the crystal lattice jostle around less, and electrons flowing through the crystal suffer fewer kicks from their vibrations.

It is therefore reasonable to expect electrical resistance to decrease as temperature is lowered. And this is indeed what happens. In the early years of the twentieth century, physicists expected that electrical resistance would gradually decrease until at absolute zero what remained would be a residual resistance produced by the electrons scattering from impurities and imperfections in the atomic lattice, and from their occasional encounters with other electrons. In order to test these ideas, Kamerlingh Onnes began a programme in 1911 to systematically measure the electrical resistance of metals at extremely low temperatures. The first measurements, performed on samples of gold and platinum, confirmed the established theories and revealed nothing unexpected.

The next metal to have its conductivity tested in the Leiden laboratory was mercury. Mercury was chosen because it was possible to obtain it in particularly pure samples. It freezes at 234 K, about 40 degrees below the freezing point of water, so it is a solid metal at cryogenic temperatures. When Kamerlingh Onnes tested mercury, he was astonished. As the temperature was lowered, the resistance in the mercury gradually decreased, just as it had for the gold and platinum samples. But then, abruptly at a temperature of 4.1 K, the resistance of the mercury sample vanished completely.[12] The resistance was not just small, but unmeasurable. Below this temperature, once a current had been established in the mercury it would flow for ever. Kamerlingh Onnes repeated the measurements with other metals, such as tin and lead, and he found that this behaviour was widespread at very low temperatures. Tin lost all resistance at a temperature of 3.7 K, and lead lost all resistance at 7.2 K. This was not simply a freakish anomaly of mercury. Kamerlingh Onnes had discovered an amazing new phenomenon, and he gave it the name superconduct-ivity. He was rewarded for his cryogenic research with the 1913 Nobel Prize in Physics.

It might seem reasonable that although the electrical resistance in superconductors is small, it cannot actually disappear altogether. However, superconducting currents have been monitored in the laboratory for over a year without any diminution in the current being detectable. From such experimental measurements it is possible to deduce that these currents would persist for hundreds of thousands of years at the very least. In fact, our modern theoretical understanding of superconductivity suggests that the currents would persist for much longer than the age of the universe.[13] The resistance in a superconductor really is zero.

Hundreds of different superconducting materials are now known. Each material has its own characteristic transition temperature, below which it is transformed into a superconductor. The transition to superconductivity is another change of phase comparable to the transition to superfluidity in liquid helium, the spontaneous magnetisation of iron below its Curie temperature, and the transformation of a liquid into a solid at its freezing point.

Levitation

In 1933 two German researchers, Walter Meissner and Robert Ochsenfeld, discovered a completely unexpected feature of superconductors: they will repel a magnetic field. A superconductor placed in a magnetic field will not allow the magnetic field to enter its interior. This remarkable characteristic of superconductors is usually called the Meissner effect, though it is sometimes referred to as the

Figure 66 The Meissner effect: a magnet photographed hovering above a superconductor.[14]

Meissner–Ochsenfeld effect, which is fairer, but a bit of a mouthful. This effect can be demonstrated by placing a magnet on top of a material before cooling it below its transition temperature. As soon as the temperature falls below this temperature and the material is transformed into a superconductor, the magnet will rise up and levitate above the superconductor, as shown in Figure 66.

The barrier to the magnetic field is not abrupt at the edge of the superconductor; it penetrates a short distance, but tails off very rapidly. In effect there is a skin around the superconductor in which the magnetic field rapidly falls to zero (see Figure 67). Within this skin, superconducting electric currents circulate and produce a magnetic field that completely cancels out the external magnetic field, leaving the innards of the superconductor completely field free. Typically the depth of this skin is around 100 nanometres, which is roughly a thousand times the size of an atom.

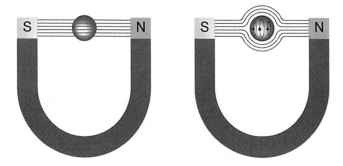

Figure 67 Schematic diagram of an ordinary conductor and a super-conductor in a magnetic field. Left: Above the critical temperature the magnetic field enters the ordinary material, represented here as a grey sphere. Right: Below the critical temperature the material becomes a superconductor, and the magnetic field is ejected from it. The black loops represent currents within the superconductor that generate a magnetic field which exactly cancels the external magnetic field inside the superconductor

The Meissner effect is remarkable to see: currents within the superconductor adjust themselves almost instantaneously, so a small magnet will hover over the superconductor. The magnet can be spun rapidly or bounced up and down, and the currents within the superconductor will automatically adjust themselves to expel the magnetic field and maintain the levitation of the magnet above the superconductor.

Super Technology

The enormous infrastructure costs of establishing a super-conducting National Grid means that it has not yet been realised. Indeed, the benefits may never outweigh the additional maintenance costs of the cryogenic refrigeration it would require. However, there are many other technological applications of superconductivity under development that may lead to the establishment of major industries that will have a great impact on our lives. Superconducting magnets have already spawned one very important industry. As there is no resistance to electric currents within a superconductor, it is very energy efficient to produce a strong magnetic field in this way. The most widespread use of these magnets, and arguably the most dramatic, is in the life-saving technology of magnetic resonance imaging (MRI), which is now routinely used to scan patients in hospitals. The superconducting magnets in MRI scanners produce magnetic fields hundreds of thousands of times as intense as the Earth's magnetic field and operate at temperatures within ten degrees of absolute zero. There are now around ten thousand MRI scanners in hospitals around the world, most of which utilise superconducting magnets.

These magnets are also at the heart of Japanese efforts to develop superfast passenger transport. Since the 1970s, the Japanese government has invested heavily in research into the technology for maglev (magnetic levitation) trains. The trains have not yet entered commercial operation, but they run along the Yamanishi maglev test line, which is 43 kilometres long and links Sakaigawa and Akiyama in Japan's Yamanishi prefecture. The Central Japan Rail Company is aiming for a commercial maglev service between the cities of Tokyo and Nagoya, a distance of around 260 kilometres, by the year 2025. The Japanese maglev trains have wheels, but the wheels retract when the train reaches high speed so that the train hovers above the guideway, levitated by the repulsion between the magnetic field of the superconducting magnets on board the train and the magnetic field of the guideway. There is no friction on the trains other than air resistance, so they are able to achieve high speeds with low energy consumption. On 2 December 2003, one of these maglev trains reached a speed of 581 kilometres per hour – a world record for a vehicle running on tracks and carrying people. Unfortunately, the economic gains of superconducting magnets are partially offset by the requirement of complicated and expensive liquid helium cooling systems around the superconducting magnets. One of the dreams of today's physicists is to find a material that superconducts at room temperature.

The Woodstock of Physics

Before 1986, the highest known transition temperature of any superconductor was just 23 K, for a compound of niobium and germanium with the chemical formula Nb_3Ge. In that year the physics community was stunned by the discovery of a compound that was superconducting up to a temperature of 38 K. It was a compound of lanthanum, barium, copper and oxygen, with quite a complicated crystal structure known as a perovskite.[15] This unexpected discovery, by Georg Bednorz and Alexander Müller at the IBM Zurich Research Laboratories in Switzerland, triggered a flurry of research in laboratories around the world. Within months, researchers found other compounds with a similar composition, but with even higher transition temperatures. A compound of the elements yttrium, barium, copper and oxygen, with the chemical formula $YBa_2Cu_3O_7$, was discovered to have a transition temperature as high as 92 K, which was a remarkable breakthrough because this material superconducts at temperatures above the boiling point of liquid nitrogen (77 K). This is extremely important with regard to technological applications, because it means that superconductors made from this material can be cooled with relatively cheap liquid nitrogen rather than expensive liquid helium. The new record transition temperature of 92 K was four times the transition temperature of the niobium–germanium compound that had held the record just months before. Even more significantly, the new record was almost one-third of the way to room temperature, which is conventionally given as 293 K, equivalent to about 20 °C. The world had entered the era of 'high-temperature superconductivity'.

At the first major physics conference after these discoveries, the 1987 American Physical Society March Meeting held in New York City, a special evening session was devoted to this sensational breakthrough. The meeting was packed with hundreds of delegates, some sitting in the gangways, while others watched the proceedings on TV screens in the hallways. The number of speakers was so great that the sessions lasted all through the night until the following morning, when the next official session of the conference was due to commence. The next day's headlines of the *New York Times* called the event 'a Woodstock for physics'.[16] Bednorz and Müller received the 1987 Nobel Prize in Physics for their breakthrough: one of the fastest Nobel awards ever.

Since the time of Bednorz and Müller's breakthrough, great efforts have been made to find compounds that exhibit superconductivity at ever higher temperatures. As of July 2012, the widely accepted record holder is a compound of mercury, barium, calcium, copper and oxygen, with a transition temperature in the region of 135 K. However, there have been claims that superconductivity has been detected above the freezing point of water in tiny samples of certain materials when subjected to ultra-high pressures.[17]

The huge technological promise of high-temperature superconductors has not yet been fulfilled. These materials are ceramics, so they are similar to pottery in structure and not easily fabricated into wires and the other components used in the electrical and electromagnet industries. However, the record for the highest magnetic field achieved by a superconducting magnet is now held by one of the high-temperature superconductors, the ceramic yttrium–barium–copper oxide. With materials such as these, commercial exploitation of high-temperature superconductivity may not be too far away. Applications in wind turbines are currently being explored. The main obstacle is the cost of a liquid-nitrogen cryogenic system to maintain the wires at a cool $-180\,°C$ (93 K). Nevertheless, we may see the first such wind farms some time in the next decade.

Cold War Science

Superconductivity remained a complete mystery to physicists for many years. But the year 1950 saw the publication of the Ginzburg–Landau theory of superconductivity. This application of Landau's theory of phase transitions was a major triumph. One of its outstanding successes was that it explained why there are two fundamentally different categories of superconductor. Most of the pure metals that superconduct are Type I superconductors – ones that lose their superconductivity when placed in quite a weak magnetic field. But Type II superconductors remain superconducting even in very strong magnetic fields, making them suitable for use as electromagnets.

The Ginzburg–Landau theory of superconductivity is still the most useful theory for analysing the results of superconductivity experiments. However, it was published in a Russian journal during the depths of the Cold War in the final years of Stalinism, a period when communications between Western scientists and their counterparts in the Soviet Union were severely hampered. For this reason, the theory received a lukewarm reception in the West. It is also true that, as the theory is based on very general principles, it did not explain the mechanism behind superconductivity or address the question of what is so special about superconductors at the level of individual particles. A few years later, in 1957, three Americans – John Bardeen, Leon Cooper and Robert Schrieffer (collectively known as BCS) – published an alternative theory of superconductivity that was received with much greater enthusiasm in the West. Whereas the Ginzburg–Landau theory is a top-down theory based on very general physical principles, the BCS theory is a bottom-up theory built around particle interactions.

The three authors of the BCS theory shared the 1972 Nobel Prize in Physics. The award of a Nobel Prize is always an outstanding achievement, but for John Bardeen this was a unique accolade as it was the second time he had won it. His first Nobel came in 1956, during the period when he was working with Cooper and Schrieffer to explain superconductivity. Bardeen shared the first award with William Shockley and Walter Brattain for the invention of the transistor, the fundamental component of computer chips found in all modern electronic devices. Bardeen remains the only person ever to receive two Nobel Prizes in Physics.[18]

At room temperature, the electrons in a metal behave as independent particles moving through a regular array of oscillating atoms. But according to the BCS theory, when a superconductor plunges below its transition temperature the electrons no longer behave like distinct individual particles. Their motion is correlated with that of other electrons. At these low temperatures, electrons with opposite spin and opposite momentum pair up, and are effectively bound together into particles called Cooper pairs. This is a bit like a proton and an electron binding together to form a hydrogen atom, but there is one obvious difference: electrons are all negatively charged, so there is a strong repulsive force between them. It therefore seems very strange that an overall attractive force could possibly hold together the two electrons forming a Cooper pair. In empty space, two electrons will always repel each other, and it is impossible for them to bind together. But when they are embedded within a lattice of positively charged atoms, it is possible for the lattice to mediate a force between two electrons such that there is a net attractive force between them. This is a strange, counterintuitive quantum effect.

QED explains the electromagnetic force as the exchange of photons between charged particles. For instance, a hydrogen atom is held together by photons that are continually being exchanged between the proton and the electron that form the atom. Similarly, according to the BCS theory, the Cooper pairs are bound together by the exchange of pulses of lattice vibrations that are passed back and forth between the two electrons. Although this binding is very weak, at very low temperatures it is sufficient to prevent the pairs from being shaken apart by the thermal vibrations of the atoms in the material, which grow weaker as the temperature falls. The result is that the Cooper pairs can flow through the superconductor completely unimpeded – there is no electrical resistance. This is similar to hydrogen atoms bathed in a sea of low-energy photons, none of which have sufficient energy to bump the electrons in the atoms up to the next rung of the energy ladder. The photons will pass through the hydrogen atoms without being able to interact with their electrons at all, and so the hydrogen gas will be perfectly transparent to the light. Once an electrical current has been established in a superconductor, the flow of the Cooper pairs that are producing the current will continue, unhindered, indefinitely. In a sense, a superconductor is transparent to the flow of Cooper pairs. If the material is warmed above its transition temperature, then the thermal vibrations of the atoms will break up the Cooper pairs and the electrical resistance will return.

These esoteric Cooper pairs may appear to defy common sense, but their existence has been confirmed in the laboratory. It has been demonstrated experimentally that the electric current in a superconductor is carried by particles that have a charge that is twice the charge of an electron. The carriers of the electric current really are Cooper pairs, not individual electrons.

A very surprising feature of Cooper pairs is that the two electrons in a pair are separated by a large distance on atomic length scales. There are hundreds of atoms in the lattice between the two electrons forming a Cooper pair. So, in any region of a superconductor there will be millions of overlapping Cooper pairs. Although the two electrons in a Cooper pair are not held tightly together, the motion of the two electrons is closely coordinated. Robert Schrieffer, the 'S' of the BCS team that cracked the mystery, likened superconductivity to quantum choreography describing the Cooper pairs in terms of a dance craze of the 1950s called the frug (pronounced 'froog'). The partners in the dance may recede to distant parts of the dancefloor, but their motions are closely coordinated even though other dancers may dance between them. Perhaps the two electrons in a Cooper pair should be thought of like two lovers who work in separate parts of a great city. Their lives are coordinated and their thoughts are interwoven, though they spend much of their time far apart mingling with the multitudes.

The BCS theory proved to be a very powerful tool for explaining the results of superconductor experiments. At first, though, there was some doubt about whether it was an acceptable theory, because it implied that the symmetry of the electromagnetic force was broken within a superconductor. This was considered to be a serious shortcoming because the symmetry is an essential part of the theory of electromagnetism – it guarantees that electric charge will be conserved in all interactions. Without this symmetry, the combination of QED and the BCS theory seemed to make no mathematical sense and would be in serious conflict with experiment. However, the relationship to symmetry breaking was clarified when the Russian physicist Lev Gor'kov proved that although the Ginzburg–Landau theory and the BCS theory look very different, they are completely equivalent at temperatures close to the transition temperature. (Among other things, Gor'kov showed that Landau's order parameter measures the density of the Cooper pairs in the superconductor.) The fact that the two theories are so different is very significant. They offer complementary views of the same physics, and this gives theorists a much better understanding of what is going on within a superconductor.

It turns out that the symmetry of the electromagnetic force really is spontaneously broken within a superconductor, and this is why a magnetic field cannot penetrate deep within a superconductor, as shown by the Meissner effect.[19] The Ginzburg–Landau theory provided a clear explanation of how the symmetry breaking arises. In fact, all electromagnetic interactions become short-range within the superconductor, not just magnetism. The Meissner effect is simply the most obvious example of how the electromagnetic forces are completely damped out over a very short distance. Inside the superconductor the electromagnetic force really is a short-range force. The interactions between the photons mediating electromagnetism and the slippery Cooper pairs changes the character of the photons. They become sluggish, as though they were wading through syrup. Effectively they have become massive particles, and this transforms the electromagnetic force. Instead of being an infinite-range force mediated by massless photons, the electromagnetic force now has a range equal to the penetration depth of the superconductor, and it is mediated by photons that behave like massive particles. Although this implies that the symmetry of the electromagnetic force is spontaneously broken within a superconductor, the symmetry is just hidden and not destroyed, so electric charge is still conserved, and the theory of electromagnetism still makes complete sense. The magic of symmetry breaking as revealed in a superconductor would now inspire the most spectacular of all the applications of Landau's theory of phase transitions: the unification of the forces of nature in the white heat of the Big Bang.

A Flaw in the Yang–Mills Jewel?

By the 1950s it was clear that QED provided an accurate and well-understood account of the electromagnetic force in terms of particle interactions. It was natural that physicists should seek to capitalise on this success and use QED as a template to model the other forces. Symmetry plays a crucial role in QED, and it almost completely determines the mathematical structure of the theory. This is also the feature of the theory that could most naturally be extended to other forces. As we saw in the previous chapter, Yang and Mills took this step and showed how to construct theories that are similar to QED, but based on larger symmetry groups. A key feature of QED is that the photon is a massless particle. This means that it always travels at the speed of light and cannot be brought to a halt. Similarly, Yang–Mills theories are mediated by collections of particles that are necessarily massless.

In the late 1950s, Julian Schwinger and his research student Sheldon Glashow proposed models with enlarged symmetries of the type devised by Yang and Mills that incorporated both electromagnetism and the weak force, and thus explained them as separate manifestations of a single force – the electroweak force. In Glashow's model the unified force was produced by the exchange of a quartet of particles – the well-established photon of QED together with a triplet of new particles: the W-plus (W^+), the W-minus (W^-) and the Z-nought (Z^0). The superscripts indicate the electric charge of each particle. The W^+ and W^- get their names from the weak force, and the Z^0 gets its name because it is uncharged: the Z is for zero. This was an elegant idea and in keeping with the dream of total unification of all the forces. However, the model had a serious drawback that effectively rendered it useless. The only way that the weakness and short range of the weak force could be accounted for was to postulate that the W-minus, the

W-plus and the Z-nought were very heavy particles. QED and its Yang–Mills generalisations have beautiful mathematical structures with elegant symmetries at their heart. Were the symmetry to be broken, the priceless Yang–Mills jewel would have a deep flaw at its centre, thus rendering it worthless. Unfortunately, the fact that the exchange particles in Glashow's theory were massive appeared to completely ruin the symmetry and thereby destroy the theory. It implied that the charges of the weak force would not be conserved, so the theory would not be mathematically consistent. The electroweak theory looked like an idea that was destined for the theoretical-physics dustbin. However, the theory was about to be rescued by a set of ideas from a most unlikely source – the physics of the ultra-cold.

The Higgs Force

Peter Higgs was born in 1929 in Newcastle-upon-Tyne in the North-East of England. His family later moved to Bristol, and he became a pupil at the same school, Cotham Grammar School, that Paul Dirac had attended. Higgs went on to attend the City of London School and then to university in London, where he completed his PhD in 1955. Five years later he moved to the University of Edinburgh, where he would eventually be appointed professor of theoretical physics. In 1964, he recognised that what particle physics required was a symmetry-breaking mechanism that hid some of the electroweak symmetry, while retaining all the nice features that symmetry brought to the theory (see Plate 9).[20]

Higgs realised that Landau's model would fit the bill nicely. By translating Landau's ideas into the language of particle physics, he constructed a model that showed how, in principle, the symmetry of a force such as electromagnetism or the weak force could be spontaneously broken. So how does the model work? Higgs invented a hypothetical field, now known as the Higgs field, that would produce a new kind of force. Associated with the field would be the particle we know today as the Higgs. The crucial feature of the model was that the Higgs field would behave like the magnetic fields of the atoms in a ferromagnet. Above a transition temperature, the lowest energy state would be one in which the Higgs field is zero. But because the Higgs field was devised in accordance with Landau's model of a ferromagnet, below the transition temperature the lowest energy state was one in which the Higgs field was non-zero, even in totally empty space. This was completely analogous to the non-zero magnetic field in a ferromagnet below the transition temperature. Below the transition temperature, the symmetry of the Higgs field would be broken.

That in itself is not so significant – who cares whether a hypothetical symmetry of a hypothetical particle is broken or not? However, what certainly did matter was the knock-on effect this symmetry breaking could have on the other forces that the Higgs particle interacted with. If the Higgs particle carried the charges of other forces, such as electromagnetism or the weak force, then the symmetry of these forces would also be broken. Now, giving the Higgs an electric charge would result in the spontaneous breaking of the electromagnetic force. The photon would become a massive particle, and electromagnetism would become a short-range force. This is not how our universe works, so it cannot be an accurate model of reality, but this was the illustrative example that Higgs presented in his original paper, and it is very similar to what actually occurs within a superconductor. The Higgs particle would be playing the same role that the Cooper pairs play in a superconductor. (For more details about the Higgs field, see the Appendix.)

Electroweak Unification

> The introduction of the idea of symmetry breaking into elementary particle physics can be compared to the breaking of the spheres by Copernicus and Kepler. It totally transformed the way of describing and understanding elementary particle interactions.
>
> Silvan Schweber (1997)[21]

Steven Weinberg was a schoolmate of Sheldon Glashow. Both Weinberg and Glashow graduated from the Bronx High School for Science in 1950 and enrolled at Cornell University to study physics. After completing his degree, Glashow went to Harvard to study for a PhD under the supervision of Julian Schwinger, while Weinberg left for the Niels Bohr Institute in Copenhagen, later returning to Princeton University to complete his PhD. Although Weinberg and Glashow did not directly work together, they would go on to share the 1979 Nobel Prize in Physics.

Weinberg was enchanted by the idea of spontaneous symmetry breaking and its potential to explain the weak force. Spontaneous symmetry breaking would ensure that all the features that follow from symmetry, such as charge conservation, which make the mathematics work and enable the theory to agree with experiment are retained. But some of the symmetry may be hidden so that it is not immediately apparent in our naive low-temperature view of the world. Higgs had shown in principle how this could work, but he had not applied his idea to a realistic model that might accurately describe the physics of the real world. By 1967, Weinberg had realised that by incorporating the Higgs field into the electroweak model devised by his old pal Sheldon Glashow, he could almost miraculously solve the problems that plagued that model. The Higgs mechanism would enable Weinberg to produce a theory that described forces that were mediated by massive particles. Although the full symmetry would not be apparent, it would still be there under the surface and this would ensure that all the crucial properties that relied on the symmetry would be retained.

The result was a unified theory of the electromagnetic and weak forces. The theory implied that these two apparently very different forces could now be considered as two manifestations of a single electroweak force, thus reducing the number of independent forces from four to three. This represents the greatest unification of the forces since the unification of the electric and magnetic forces by Faraday and Maxwell in the nineteenth century. The theory is known as the Glashow–Weinberg–Salam (GWS) model. It was for developing this unified electroweak theory that Weinberg and Glashow were awarded their Nobel Prize in Physics, along with a third recipient, the Pakistani physicist Abdus Salam, who was also exploring similar models in the 1960s.

The electroweak theory implies the existence of a new force – the Higgs force – that acts on most other particles and affects their properties.[22] In this theory the Higgs particle has no electric charge, but it does carry the charges of the weak force, so it interacts with the mediators of the weak force, but not the photon. According to the theory, in the earliest moments of the universe the symmetry of the electroweak force was unbroken and all four of the particles mediating the force would have been massless. An instant later, as the temperature fell below about a trillion degrees, the universe underwent a dramatic phase transition. Since this time the Higgs field has had a non-zero value everywhere, even throughout empty space. The result is that the electroweak force is broken. Only one of the force mediators – the photon – remains massless, because it does not interact with the Higgs, while the other three – the W-plus, W-minus and the Z-nought – become very heavy because of their interactions with the Higgs field. As the W and Z particles move through this background field of the Higgs, they are tugged back by the Higgs force, and this is what gives them their large mass. The result is that the electromagnetic force mediated by the photon is a powerful long-range force, whereas the weak force mediated by the other three particles is a feeble and extremely short-range force. The Higgs particle plays the same role in empty space that the Cooper pairs play in a superconductor. To paraphrase the Nobel Prize-winning physicist Frank Wilczek, we are all living within a 'cosmic superconductor'.[23]

We might wonder whether space really can be filled with the Higgs field without us being aware of its presence. But we are surrounded by electromagnetic fields, such the Earth's magnetic field and all the electromagnetic fields produced by our televisions and other electrical devices, and we are not consciously aware of these fields, although the electrically charged particles within us must be affected by them. Similarly, as we move through the background Higgs field it also affects the particles from which we are composed. According to the electroweak theory, the Higgs force does not just give mass to the W and Z particles, but is also responsible for the mass of the fundamental matter particles, such as the electron and the quarks. These matter particles would also be massless if it were not for their interaction with the Higgs field.[24] Arguably, we have a much greater physical awareness of the presence of the Higgs field than we do of electromagnetic fields, because the Higgs field gives rise to the mass of our subatomic components – and we are, after all, conscious of our weight.

If the Higgs field is all around us, why are we not surrounded by Higgs particles in the same way that we are bathed in photons? The explanation is that photons are massless, so it takes very little energy to produce a photon, and consequently we are surrounded by photons all the time. By contrast, the Higgs is an extremely heavy and unstable particle, so although we are constantly moving through the background Higgs field, there are no Higgs particles around. Higgs particles would have been present in the very early universe, but as the temperature dropped they would have quickly decayed into lighter particles. They have all long since disappeared. This is why it is necessary to construct an enormous particle accelerator to create the conditions in which the Higgs particles will be produced. This is the principle motivation for building the Large Hadron Collider at CERN. Superconductivity is playing a remarkable dual role in this effort. Not only is it the theoretical inspiration behind the theory of the Higgs particle, but it is also responsible for the cutting-edge technology of the machine. The ultra-high-energy protons are guided around the 27-kilometre circuit of the collider by the intense magnetic fields of niobium–titanium superconducting magnets cooled in a bath of super-fluid helium.

We should not forget that, although theorists such as Peter Higgs and Steven Weinberg might build models in which hypothetical particles like the W-minus, W-plus and Z-nought, or the Higgs, behave in a particular way, the truth depends entirely on whether the theory provides an accurate description of the real universe. In 1967 it was not at all clear that the electromagnetic and weak forces could be unified in the way described by the electroweak theory. This could only be decided by conducting experiments to test the theory's predictions. As the next chapter will show, the theory is a spectacular success: so far it has passed every test with flying colours.

Chapter Eight

THE GRAND SYNTHESIS

I have always felt that science, technology, and art are importantly connected; indeed, science and technology seem to many scholars to have grown out of art. In any case, in designing an accelerator I proceed very much as I do in making a sculpture. I felt that just as a theory is beautiful, so, too, is a scientific instrument – or that it should be. The lines should be graceful, the volumes balanced. I hoped that the chain of accelerators, the experiments, too, and the utilities would all be strongly but simply expressed as objects of intrinsic beauty. Aesthetics is partly a matter of communication, and with so many people involved, I felt that everyone would appreciate the economy of good design and would keep their designs equally clean and *understood*.

Robert R. Wilson, Director of Fermilab (1987)[1]

The Lord of the Rings

Robert Rathbun Wilson was born on his family's cattle ranch in Wyoming in a town called Frontier on 4 March 1914. As a youngster he lived close to nature and learned the skills of a horseman on the wild American frontier. His youthful aspirations were simply to be a cowboy. Not surprisingly, his school record was mediocre, but he managed to win a place at the University of California at Berkeley, which he entered in 1932. As a new student, Wilson was shown around the campus. He was left uninspired until the moment he was invited into the Radiation Laboratory to see the accelerator known as the cyclotron. This machine had recently been invented by Ernest Lawrence. Instead of accelerating particles in a straight line, Lawrence's machine accelerated them in a spiral trajectory within two D-shaped magnets. The cyclotron and its successors were far more effective than the linear accelerators of the day, enabling Lawrence to accelerate protons to energies of over 1 MeV for the first time. For Wilson it was love at first sight. The cyclotron looked like a machine from the future or a science fiction movie, and he knew at once that he would dedicate his life to working with such machines. He would soon become an important member of Lawrence's team – a team that produced many of the leading figures in American post-war experimental physics. Bob Wilson, as he became known to his colleagues, would go on to design and develop many of the fundamental components of future particle accelerators.

A decade later, Wilson was invited to join the Manhattan Project, the American programme to develop nuclear weapons during the Second World War. This top-secret project was based in the small desert town of Los Alamos, New

Mexico. Wilson became head of the Experimental Nuclear Physics Division; at twenty-nine he was the youngest of the division leaders at Los Alamos. John Adams portrayed him in this role as one of the leading characters in his recent opera *Doctor Atomic* (2005), which tells the story of the first atomic bomb test. Wilson's energy and charisma also resulted in him being elected mayor of the town of Los Alamos.

After the war, Wilson joined the physics department at Harvard, where he designed and built a series of electron accelerators and also pioneered proton therapy as a treatment for cancer. There are now dozens of proton therapy centres in operation around the world, each of which requires a powerful particle accelerator to produce the high-energy protons. The use of various forms of radiation to treat cancer has become routine. The largest particle accelerators in the world may be used for fundamental physics research, but in total over 95 per cent of all these machines are now found in hospitals.

By the early 1960s, the two leading American particle physics laboratories were based at Berkeley, California, and Brookhaven, New York. There was a great rivalry between these two institutions on opposite coasts of the United States. But particle physics was becoming progressively more expensive, coming to rely increasingly on ever larger and more powerful machines. After much wrangling, a decision was taken that the next generation particle accelerator should be a national facility that all American physicists could use. This instrument would accelerate protons to 200 GeV – five times the energy attainable by any existing accelerator – before smashing them into a fixed target.

While at a conference in Frascati in Italy in September 1965, Wilson attended a presentation of the preliminary designs for the new American accelerator by a member of the group from Berkeley. Wilson, not impressed, was quick to express his dismay. He felt that a design energy of 200 GeV was much too conservative. With the $250 million allocated to the project, he believed that a target of 400 GeV or even 500 GeV was achievable, and that the challenge of such an ambitious goal would attract the talented people who were needed to ensure the success of the endeavour. After the conference, Wilson headed off to Paris where he had enrolled for a fine art course, but he couldn't stop thinking about the accelerator project. Even during his art class, he found himself dreaming up new machines and calculating their specifications. Like the opposite of a naughty schoolboy, beneath the sensuous curves of the beautiful model he was drawing, he was furtively sketching designs for improved particle accelerators with their own elegant curves. Wilson remained in Paris for several days after his art class. 'I went around sitting in cafés,' he said. 'I spent the whole time going through one machine after another in a fury, making all kinds of designs'.[2] Back in America, he launched his campaign against the conservative Berkeley design.

The go-ahead for the construction of the accelerator finally came in 1967. Rather than favouring either of the established laboratories, the decision was taken to choose a new site in the farmlands of Illinois, close to Chicago. To the great shock of many American physicists, Bob Wilson was given the task of establishing this state-of-the-art, world-beating laboratory in the rural Midwest. Initially the laboratory was simply called the National Accelerator Laboratory, but it was later renamed the Fermi National Accelerator Laboratory in honour of the great Italian-American physicist Enrico Fermi, and is now generally known as Fermilab. Wilson was determined that the accelerator would be completed ahead of schedule and that the costs should be kept as low as possible. In his view, if anything worked first time, it was clear that it had been over-designed and was therefore unnecessarily expensive. He would rather save money and take risks with the confidence that whatever failed initially could be corrected later. As he put it, 'so that no one could charge me with "not being crazy enough", I cut the construction schedule of about seven years ahead to about five years'.[3]

Wilson moulded the design of the Fermilab site with the eyes of a sculptor modelling a figure, determined that the curves of the entire accelerator complex should meet his aesthetic ideals. The accelerator would consist of a linear accelerator that would fire the protons into a booster ring, where they would be accelerated further before entering the six-kilometre circuit of the Main Ring, in which they would achieve their maximum energy. Although the accelerators were to be built well underground, Wilson decided that their outline should be visible on the surface. At Fermilab, grassy banks and arc-shaped ponds mark out the paths of all the underground tunnels, providing a spectacular view of the accelerator complex from the air (see Plate 10). Wilson was determined to build a physicists' paradise: 'a utopian laboratory [that] clearly required a setting of environmental beauty, of architectural grandeur, of cultural splendour'.[4] He felt that the site should include just one major building, rather than having a scattered array of several smaller buildings, and that this structure should be located at the point where the protons were injected into the Main Ring. To determine the appropriate height of the building, Wilson was taken up in a helicopter to get a feeling for the aesthetics of living in the building. He decided that the sky, the sunsets and the Illinois landscape were best appreciated from a height of at least 200 feet. That helicopter flight determined that the final height of the building, known today as Wilson Hall, would be 250 feet. Its design was inspired by Beauvais cathedral in northern France, and it does indeed rise up from the prairie, dominating the landscape like a medieval cathedral.

In the grounds of Fermilab, Wilson erected a number of large abstract sculptures with names such as Broken Symmetry and Möbius Strip. Many of these artworks were constructed out of decommissioned metal components from the laboratories. Wilson even designed his own elegant electricity pylons, which he named 'power poles', to bring the power supply for the Fermilab accelerators. He initiated a programme to restore the land occupied by Fermilab to the original prairie that the early settlers would have encountered. He reintroduced native grasses and other indigenous plants, along with a herd of bison to graze the grassland. This work continues today, with volunteers from the laboratory planting seeds and tending the land. The Fermilab grasslands are now home to many rare flowers, insects and birds.

However, the result of Wilson's cost-slashing measures was that the facilities constructed for the physicists were austere. Researchers were treated like pioneers on the prairie, housed in temporary accommodation and faced with a daily battle against leaking roofs and muddy floors in the experimental areas. Their efforts were heroic – teasing data out of their apparatus while labouring in very difficult and uncomfortable conditions. But Wilson instilled in his team a frontier mentality, filling them with a pioneering spirit in which experimental physics became an adventure into the unknown.

As the Main Ring approached completion in the summer of 1971, disaster struck. The magnets that had the job of guiding the proton beams around the accelerator began to short-circuit and fail. Almost a third of the one thousand magnets failed during this period of crisis. Many of Wilson's critics expected that the folly of placing this high-profile and expensive project in the hands of such an eccentric director was about to be revealed. The start-up of the accelerator was postponed for six months as the physicists struggled throughout the winter to fix the problems with the magnets. By 21 January 1972 the frustrated physicists were still battling to produce a stable proton beam. One researcher later recalled that at a critical moment late that evening, just as despair was setting in, their enigmatic leader paid a timely visit to the control room. Wilson appeared, opened a small book and began to read. With passion in his voice, he rallied his troops with a speech that inspired the armies of Emperor Charlemagne in the Old French epic, the *Song of Roland*. Somehow Wilson's sorcery worked. The next day, as if by magic, a stable 20 GeV beam circulated in the Main Ring for the first time. By 11 February a record energy of 100 GeV had been achieved, and on 1 March the machine reached its design energy of 200 GeV. Wilson could now triumphantly announce to the world that the accelerator had been completed well under budget and years ahead of schedule. The instrument would eventually achieve Wilson's target of 500 GeV in May 1976.[5]

The Quantum Symphony

> The state of particle physics [is] not unlike the one in a symphony hall a while before the start of the concert. On the podium one will see some but not yet all of the

musicians. They are tuning up. Short brilliant passages are heard on some instruments; improvisations elsewhere; some wrong notes too. There is a sense of anticipation for the moment when the symphony starts.

Abraham Pais (1968)[6]

Abraham Pais's vision of a vibrant and melodious future for particle physics was remarkably astute. The general perception of the physics community at the time was that much of what was going on in particle accelerators was far from being understood. But Pais could discern the glimmerings of a new synthesis of the forces of nature emerging. He believed that physicists were on the verge of a fundamental breakthrough in the unification of the forces.

Indeed, most of the main parts of Pais's symphony had already been composed by one theorist or another. But many of these themes, including quarks, Yang–Mills theories and the Higgs mechanism, were not yet widely recognised as the great hit anthems they would soon become. Gell-Mann had shown how the properties of all the particles that interact via the strong force could be understood by assuming them to be composed of quarks. But even Gell-Mann himself hedged his bets about the physical reality of quarks. Julian Schwinger and Sheldon Glashow had shown how the electromagnetic and weak forces could be described by a Yang–Mills theory, an elegant type of theory that is modelled on QED, but with a larger symmetry group. This would unify the two interactions into a combined electroweak force mediated by a quartet of particles: the massless photon of electromagnetism and the W-minus, W-plus and Z particles that would mediate the weak force. Peter Higgs had discovered how the symmetry of a Yang–Mills theory could be spontaneously broken without destroying the theory, and Steven Weinberg had demonstrated how the mechanism devised by Higgs could be employed to spontaneously break the symmetry between the electromagnetic and weak forces. This would enable the W and Z particles to become massive particles, and explain the weakness and short range of the weak force. All the main components had now been devised, but it would be several years before their full significance became clear to the physics community.

Weinberg's landmark electroweak paper was published in 1967.[7] It would be natural to assume that the arrival of a theory that unified the electromagnetic and weak forces would have been acclaimed as a fantastic triumph. After all, it was over a century since James Clerk Maxwell had performed a comparable feat – the unification of the electric and magnetic forces and the creation of the theory of electromagnetism. However, the reality was very different: Weinberg's paper passed almost unnoticed. It wasn't referred to at all in any publication in the period between its first appearance in 1967 and the end of 1969. In 1970 it was cited in just one research paper; the following year it was cited in three. Even Weinberg himself seems to have forgotten about the paper, which is remarkable as it would eventually win him a Nobel Prize.[8] The reason for this astonishing neglect of such an important paper was that throughout the 1960s there were many other avenues of research that looked more promising. Most theorists concentrated on a range of exotic ideas with names like bootstrap models, the analytic S-matrix and dual-resonance models. Today these brilliant and plausible ideas of the past have been relegated to historical footnotes. It took at least a decade of further study and a great deal of effort by many theorists and experimenters for the new approach to be assimilated. Ultimately, physicists were decisively won over when the new ideas were supported spectacularly by experimental evidence. This story is one of the main themes of this chapter.

In the early 1970s, a young Dutch theoretical physicist, Gerard 't Hooft, working with his PhD supervisor Martinus Veltman, made a series of technical breakthroughs. Essentially they achieved for Yang–Mills theories the equivalent of what Feynman, Schwinger, Dyson and Tomonaga had done for QED. Their insights provided a deeper understanding of the mathematical structure of these theories and produced the tools that enabled detailed calculations to be performed. For many theorists, this was the crucial turning point. They could now see that, rather than merely being abstract curiosities, Yang–Mills theories might be the appropriate models to describe the real world of elementary particles, and this gave them the confidence to study their implications in depth. In the words of leading theorist Sidney Coleman, "'t Hooft's kiss transformed Weinberg's frog into an enchanted prince'.[9] The importance of Weinberg's electroweak paper was now apparent. In 1972 it received 65 citations, and in 1973 it received 165 citations.[10] This number would continue to rise through the rest of the decade. The work of 't Hooft and Veltman was probably even more important than the work of Weinberg, Salam and Glashow in establishing the modern synthesis of particle physics. However, it wasn't until 1999, almost thirty years later, that 't Hooft and Veltman were rewarded with the Nobel Prize in Physics.[11]

All the ingredients were now ready for a consolidation of particle physics, and over the next few years all the pieces would fall into place. The name given to this grand synthesis is the Standard Model of particle physics. This theory encompasses the whole of particle physics; in fact, our entire understanding of the workings of the universe (apart from gravitation) is encapsulated in this single theory. It is one of the truly great achievements of twentieth-century science. The Standard Model unifies electromagnetism and the weak force into the Glashow–Weinberg–Salam electroweak theory, which it combines with the strong (or colour) force as described by quantum chromodynamics (QCD). Both parts of the Standard Model are the same type of theory. They are both Yang–Mills theories, but built around different symmetry groups. The Standard Model unites all the particles from which matter is formed, as well as all the forces that hold matter together. Furthermore, it is not wild speculation; quite the opposite – it makes very precise predictions that can be tested in the laboratory. Unless these predictions hold up, then the Standard Model must be wrong. Experiments at particle accelerators provide the best way to test these predictions.

The key prediction of the electroweak theory is the existence of the massive W-plus, W-minus and Z particles that mediate the weak force. The theory also requires the existence of the Higgs particle to break the symmetry between the electromagnetic and weak forces, as described in detail in the previous chapter. The theory does not tie down the mass of the Higgs, but it makes definite predictions about the masses of the W and Z particles. When these calculations were first done, it became clear that the W and Z particles were too heavy to be generated in existing accelerators. Their discovery, and the confirmation of the electroweak theory, would require the development of new technologies and the construction of larger and more powerful accelerators than any yet conceived. This would lead to a transatlantic race to reach the highest energies, and prizes would be won on both sides of the Atlantic. Over the next two decades the world of particle physics research would come to be dominated by two great laboratories: Fermilab and CERN.

Who Ordered *That?*

The Standard Model provides a unified description of the fundamental constituents of matter. Gell-Mann had shown how the properties of all the particles that interact via the strong force could be understood if they were composed of quarks (and antiquarks). Quarks are bound together by the colour force to form particles such as protons, neutrons and pions that are collectively known as hadrons. These hadrons could no longer be considered to be elementary particles. Quarks form a deeper layer of matter from which all strongly interacting particles are formed. But, as described in Chapter 6, there are other particles, such as the electron, that do not feel the strong force. The electron cannot be composed of quarks: if it were, it would also interact via the strong force. As far as anyone knows, the electron is a fundamental par-ticle in its own right; it is not composed of more fundamental constituents. This puts the electron on a par with the quarks as an elementary component of matter. A second particle, encountered in earlier chapters, that does not feel the strong force, is the humble neutrino. The neutrino is also considered to be elementary.

There is one other particle that has not yet entered our story. It was first spotted in cosmic-ray photographs in the 1930s and is known as the muon. For a number of years the muon was confused with the pion, and the true identities of these two very different particles was not established until 1947. The muon feels the electromagnetic and weak forces, but not the strong force. In fact, the muon seems to be an exact replica of the electron in every respect except its mass.

Whereas the mass of the electron is 511 keV (0.511 MeV), the mass of the muon is 105 MeV, making it just over two hundred times as massive as the electron. It appears to be a heavyweight relative of the electron. The discovery of the muon came as a shock. It arrived in the days when the structure of matter appeared to be well understood in terms of electrons, protons and neutrons. Dirac had predicted the existence of the positron, which had recently been found by Anderson. Pauli had postulated the existence of the neutrino. But no one had anticipated the existence of a heavier version of the electron. It seemed completely unnecessary for the correct functioning of the universe. The surprise of the physics community is summed up in Isidor Rabi's rather irritated response on learning of the muon's discovery: 'Who ordered *that*?' Muons do not feel the strong force. They are not composed of quarks. As far as is known, the muon is another fundamental particle without any substructure.

In radioactive beta decay, the first manifestation of the weak force to be discovered, the emission of an electron from an atomic nucleus accompanies the emission of a neutrino which usually departs, never to be seen again. The production of muons is also associated with the emission of neutrinos. For instance, when a pion is produced by a cosmic-ray collision high in the atmosphere, or in a particle interaction in an accelerator, the pion will typically have a lifetime of about ten nanoseconds (10^{-8} seconds). It will then decay via the weak force into a muon and a neutrino:

$$\pi^- \to \mu^- + \nu_\mu$$

Here the muon is represented by the Greek letter mu (μ). The neutrino,[12] represented as usual by the Greek letter nu (ν), has a subscript mu because the muon neutrino is not the same particle as the electron neutrino. This difference was first demonstrated in an ingenious experiment by the American physicists Leon Lederman, Melvin Schwartz and Jack Steinberger at the Brookhaven National Laboratories in 1962. This trio were rewarded with the 1988 Nobel Prize in Physics.

The muon will itself decay within about a microsecond (10^{-6} seconds) in the following reaction:

$$\mu^- \rightarrow e^- + \nu_\mu + \nu_e$$

where ν_μ is the muon neutrino and ν_e is the electron neutrino.[13] Electrons, muons and their associated neutrinos are collectively known as leptons. This gives us two classes of matter particle: leptons and quarks. All matter is formed from leptons and quarks: this is the great subdivision in the composition of matter. These particles are the modern equivalents of the chemical elements, the ultimate building blocks of matter. Quarks feel the strong force – that is, the colour force described by QCD – but leptons do not feel the strong force. Both quarks and leptons feel the weak force. The weak force is responsible for the interconversion of quarks into other quarks and for the interconversion of leptons into other leptons. No known force will convert a quark into a lepton or a lepton into a quark.

A quick head count gives a total of just seven fundamental matter particles. These are four leptons, e, μ, ν_e and ν_μ, and three quarks, u, d and s (up, down and strange), along with their seven antiparticles. This small collection of fundamental particles accounts for the structure of matter in a very neat and tidy way.

Gargamelle

> In the vigour of his age he married Gargamelle, daughter
> to the King of the Parpaillons, a jolly pug, and well-
> mouthed wench. These two did oftentimes do the two-
> backed beast together, joyfully rubbing and frotting their
> bacon 'gainst one another, in so far, that at last she became
> great with child of a fair son, and went with him unto the
> eleventh month; for so long, yea longer, may a woman carry
> her great belly, especially when it is some masterpiece of
> nature, and a person predestinated to the performance,
> in his due time, of great exploits.

François Rabelais, *Gargantua and Pantagruel* (1532)[14]

Before the Second World War, Europe led the world in
theoretical physics. By the end of the war, the United States
had forged ahead. This was largely the result of the exodus of
leading physicists escaping the oppression of 1930s Europe.
Many of the physicists were Jewish, fleeing from the tyranny
of the Nazis and their allies. The American pre-eminence in
physics was confirmed by the military research conducted
during the war, and especially the success of the Manhattan
nuclear weapons project.

The establishment of a pan-European laboratory was first
proposed in 1949 by the physicist Louis de Broglie. He had
triggered the quantum revolution of the 1920s with his suggestion that matter particles, such as electrons, might have
wave-like properties. The lab was conceived in an attempt
to promote international collaboration in the years following
the devastation of the war. De Broglie also suggested that a
major European centre for physics might help to stem the
flow of European physicists to America. On 29 September

1954 the CERN convention was ratified by eleven European nations, and the new laboratory came into existence. There are now twenty member countries and several others with observer status. CERN is still known by its original acronym, even though the full name of the laboratory is now the European Organization for Nuclear Research. In contrast to the United States, where most post-war federal science research has been funded by the military,[15] CERN has always been keen to emphasise its role in the peaceful exploitation of nuclear physics. The original laboratory was built near Geneva in Switzerland, close to the French border. It has evolved into a huge accelerator complex that now crosses the border into France.

For many years, CERN lagged behind its American counterparts, but it came into its own in the quest for the origins of the electroweak force. In 1970, a gigantic new detector was installed at the laboratory. It was barrel-shaped, 4.8 metres long with a diameter of 1.85 metres, and weighed in at about 1,000 tonnes. The detector was filled with 18 tonnes of liquid freon in which the tracks of charged particles would appear as trails of tiny bubbles curving in a strong magnetic field that would enable researchers to distinguish between positively and negatively charged particles. The enormous bubble chamber had been built at the French physics laboratory at Saclay near Paris, where it was named Gargamelle after the heavy-drinking, tripe-eating, vulgar mother of the giant Gargantua – the eponymous hero of Rabelais's bawdy sixteenth-century novel. Gargamelle would play a major role in gathering evidence for the electroweak theory of Glashow, Weinberg and Salam. As discussed in the previous chapter, an important consequence of uniting electromagnetism and the weak force in this theory is that the electroweak force is mediated by a quartet of particles: the photon that mediates the electromagnetic force and the W-minus, W-plus and Z-nought that mediate the weak force. The effects of the exchange of the W-minus and W-plus particles were already familiar to physicists. If the electroweak theory was correct, radioactive beta decay and the transmutation of the elements in the stars were caused by the exchange of W particles.

In all the weak interactions that had ever been observed, the identities of the particles undergoing the interaction would change: for example, neutrons would change into protons, or protons into neutrons. (At the quark level this corresponds to a down quark changing into an up quark, or vice versa.) According to the electroweak theory, the exchange of

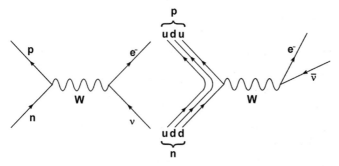

Figure 68 Feynman diagrams showing a weak interaction mediated by a W particle. Left: This interaction corresponds to beta decay, which occurs in an atomic nucleus when a neutron is converted into a proton with the release of an electron and a neutrino. The nucleus gains another proton, so the atom is transformed into a new type of atom. This transmutation of elements occurs in the stars. Right: The process can be understood at a deeper level in terms of quarks. The neutron is composed of two down quarks and one up quark, whereas the proton is composed of two up quarks and a down quark. Beta decay occurs when one of the down quarks in the neutron converts into an up quark and emits a W-minus particle. The W-minus then decomposes into an electron and a neutrino.

W particles would explain how this occurs, as shown in the Feynman diagram in Figure 68. But the effect of exchanging a Z particle would be different: it would be more like the exchange of a photon, and this had never been observed in the laboratory.

The electromagnetic force between two charged particles is produced by the exchange of photons between the charged particles. As illustrated in Figure 69, the electroweak theory of Glashow, Weinberg and Salam predicted

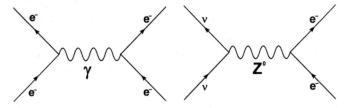

Figure 69 Left: The electromagnetic repulsion between two electrons is caused by the exchange of photons between the electrons. Right: The electroweak theory of Glashow, Weinberg and Salam predicted that there should be a similar force between neutrinos and electrons produced by the exchange of the Z particle.

that there should be a similar force due to the exchange of the Z particle. Physicists refer to these interactions as weak neutral currents, because they are weak interactions that are mediated by the neutral, or uncharged, Z. One manifestation of a weak neutral current would be a hitherto unobserved force between neutrinos and electrons. Of course, neutrinos are famous for being elusive, so observing an interaction in which a neutrino hits an electron and knocks it out of an atom would be extremely difficult. In 1973, researchers at CERN took up the challenge of identifying this effect.

CERN's first major accelerator, the Proton Synchrotron, had been built in the 1950s to accelerate protons up to an energy of 28 GeV. In the search for weak neutral currents, intense beams of muon neutrinos[16] were produced by the Proton Synchrotron and fired into the Gargamelle bubble chamber. The aim was to see the track of an electron racing through Gargamelle after being struck in an extremely rare encounter with a muon neutrino. In July 1973, scientists at CERN announced that their search had been successful. They had scanned almost three-quarters of a million photographs and found one that showed exactly the type of event they were looking for. A single example is certainly not conclusive evidence for a new type of interaction, but the discovery was later confirmed by further observations at both CERN and Fermilab. Weak interactions produced by the exchange of the Z particle really were being observed, just as the electroweak theory predicted. This was the first big discovery at CERN and the first time that a prediction of the Glashow–Weinberg–Salam electroweak theory had been substantiated in the laboratory.

Charm Offensive

> I do not claim any deep understanding of *Finnegans Wake*, but I believe that, had Murray Gell-Mann known the existence of more than three elementary constituents of hadronic matter, he would have chosen a different name.
>
> John Iliopoulos (1987)[17]

In 1970, Sheldon Glashow, John Iliopoulos and Luciano Maiani, whose collaboration is often referred to by the acronym formed from their initials (GIM, but pronounced 'Jim'), argued that some puzzling experimental results could be explained if there were a fourth type of quark. Gell-Mann had originally proposed the existence of three flavours of quark: the up quark, the down quark and the strange quark. The up quark carries an electric charge of $+\frac{2}{3}$, whereas both the down quark and the strange quark carry an electric charge of $-\frac{1}{3}$. GIM proposed that their new quark must have an electric charge of $+\frac{2}{3}$, like the up quark. This would mean that the new quark was like a heavy version of the up quark, in the same way that the strange quark is like a heavy version of the down quark.

Glashow gave the hypothetical new quark the name 'charm', as it seemed such an elegant way to explain the riddle posed by the experimental data.[18] The quarks and leptons would then fall neatly into two generations of four particles, each generation consisting of two leptons and two quarks. The first generation would contain the electron and the electron neutrino, and the down and up quarks. The second generation would consist of the muon, the muon neutrino and the strange and charm quarks. Each of the four particles in the second generation would carry exactly the same charges as its lighter relative in the first generation, so they would behave in the same way in electromagnetic, weak and strong interactions. The second generation was therefore an exact replica of the first generation, but with much heavier particles.[19] This was the first hint of a new pattern in nature. According to Glashow, the fundamental matter particles seemed to be ordered into repeated sets of four particles.

Quark, Strangeness and Charm

> What a lady looks for in her lover
> Is charm, strangeness and quark.

Bob Calvert and Dave Brock (Hawkwind),
'Quark, Strangeness and Charm' (1977)

In the early 1970s there were still many doubts about the existence of quarks, but this would all change in 1974. On 4 November of that year, Burton Richter, the leader of a team at the Stanford Linear Accelerator (SLAC) in California, and Sam Ting, a team leader at Brookhaven, simultaneously announced the discovery of a new particle. Both had found a very strong signal for the new particle, but using completely different detector systems at very different accelerators. The Brookhaven team called the particle J, because the letter J looks similar to a Chinese character that is pronounced 'ting', and the SLAC team named the particle after the Greek letter ψ (pronounced 'psi'). Uniquely, this particle acquired a double-barrelled name and is still referred to as the J/ψ (jay-psi). Glashow immediately claimed the discovery as the first sighting of his fourth quark, charm. He believed that the J/ψ was a meson formed of a charm quark bound to an anticharm quark – and he was soon proved to be correct. His interpretation was backed up by the discovery that the J/ψ was only the first member of a family of mesons with similar masses. The reason for the existence of a family of such particles is that when the charm quark and the anticharm quark are bound together, they can whirl around each other in a number of different energy levels, just like an electron in a hydrogen atom. Because of the equivalence of energy and mass, each step up the energy ladder for the charm–anticharm pair corresponds to a meson of higher mass. The name that is given to the charm–anticharm system is 'charmonium'.

There are currently at least nine known mesons in the charmonium system. Of these, the J/ψ, with a mass of

3,097 MeV, is the second lightest and the longest lived. Another example of a charmonium meson is ψ(2S), whose mass is 3,686 MeV – almost 600 MeV heavier than the J/ψ. The mesons with a higher mass usually decay rapidly into a J/ψ meson with the release of a photon, just like an electron falling down the energy ladder in a hydrogen atom. Alternatively, heavier mesons might decay into a J/ψ with the emission of a couple of pions. The J/ψ then decays into various lighter mesons formed of the much lighter up, down and strange quarks.

Glashow's arguments convinced the physics community. The discovery of charm was heralded as the greatest discovery in particle physics since the discovery of the first 'strange' particles in 1947. The importance of quarks could no longer be denied. Now everyone was a believer, and a new epoch in particle physics had begun. Physicists remember the overthrow of the old order as the November Revolution of 1974.

It soon became apparent that the existence of the fourth quark was even more important than had been suggested in the original GIM paper. Theorists now realised that the structure of the Standard Model requires matter particles to form complete generations of four particles. With incomplete generations, the Standard Model would not be a viable theory. The consistency of Yang–Mills theories is a delicate balancing act. It is possible for symmetries that exist in a classical theory to be destroyed in the quantum version of the theory. For rather technical reasons, this is what would happen if there were an incomplete generation of particles, but this cannot be allowed to happen, because without the symmetry the theory is useless. If the symmetry were lost, the theory would also lose its mathematical consistency, and it certainly couldn't then describe the forces of nature.

Fortunately, the electroweak and colour charges of the four particles in each generation match up in exactly the right way to cancel out these potential problems. Furthermore, these cancellations demand the existence of three colours of quarks – no more, no less, just as QCD implies. So it was now clear that all the components of the Standard Model – the electroweak theory, QCD and the sets of four matter particles that form each generation – mesh together perfectly to make a consistent theory. Each part of the Standard Model was devised separately to fit the experimental facts, but the parts fit together like a hand in a glove. Surely this was no accident.

Most combinations of matter particles and forces would not produce a consistent theory. But the Standard Model fits together harmoniously, suggesting that it really does accurately represent how the universe works at a deep level. The Standard Model is a grand synthesis of the fundamental forces and particles. It combines the whole of particle physics together into a single elegant theory. The discovery of charm was a triumph for the Standard Model, as it seemed to wrap everything up nicely. Two complete generations of fundamental particles were now known: four flavours of lepton and four flavours of quark.

Beautiful Bottoms

> 'Beauty is truth, truth beauty,' – that is all
> Ye need on earth, and all ye need to know.
>
> John Keats, 'Ode on a Grecian Urn' (1819)

In the late 1960s, Martin Perl was part of a team investigating whether there were any differences in the way that electrons and muons interact. The outcome of these

experiments was confirmation that the strength of the interactions of these two particles are exactly the same, and that
the muon is indeed just like a very heavy electron. By 1971,
Perl was speculating about a third member of the family, a
particle that behaved like the electron and the muon but was
even heavier than the muon. Perl and his colleagues wrote
proposals for how such a particle could be tracked down at
the new electron–positron collider SPEAR that was due to
be commissioned at the Stanford National Accelerator Laboratory in California in 1973.[20] During the first two years of
its operation, Perl amassed data from the new collider, and
by December 1975 he was sufficiently confident to suggest
that they were observing the signal of a new particle with
a mass between 1.6 and 2.0 GeV. However, this conclusion
was rather tentative. The energy for producing a pair of the
new leptons was close to the energy region of the J/ψ and
its charmonium relatives, so it took a while to disentangle
the data. There was also the question that, although there
had been good theoretical reasons for there to be a fourth
quark to complete the second generation of particles, there
was no compelling reason for there to be a new heavyweight
relative of the electron and the muon. It would take several
years of gathering data to demonstrate conclusively that the
new particle was decaying in the manner expected of a new
heavy lepton and to confirm that this was indeed what had
been discovered. The new particle was given the symbol τ,
the Greek letter tau (which rhymes with 'now'), because this
is the initial letter of the Greek word triton, which means
'third'.[21] This third generation heavy lepton is now known as
the 'tauon'. Its mass is a whopping 1.78 GeV. By comparison,
the mass of the electron is 511 keV, and the mass of the muon
is 105 MeV, so whereas the muon is about 200 times the mass

of the electron, the tauon is almost 3,500 times the mass of the electron. In 1995, Martin Perl was awarded the Nobel Prize in Physics for this discovery.

No sooner had the charm quark completed the second generation than the discovery of the tauon implied the existence of a full third generation of matter particles. The consistency of the Standard Model demands that matter particles exist as complete generations, so if the Standard Model was to be a reliable account of particle physics, there had to be two more flavours of quark awaiting discovery, as well as a third neutrino, the tauon neutrino. The two missing quarks were named top and bottom, to reflect the fact that they were the third generation equivalents of the up and down quarks. Often the names of the quarks were abbreviated to their initials 't' and 'b', and some physicists began to refer to them as the truth and beauty quarks, continuing the poetic terminology of strangeness and charm. Relating beauty to truth seems very apt to modern physicists, as their most beautiful theories seem to produce the truest reflections of reality. Unfortunately, it is the more prosaic 'top' and 'bottom' terminology that is used most frequently today.

Physicists did not have to wait long for the first of these two quarks to make its appearance. Leon Lederman, the leader of one of the teams at Wilson's Fermilab, had narrowly missed the discovery of charm. Lederman felt that there had been three occasions during his career on which he might have discovered the charm quark, but it had eluded him each time. His team was determined to press on and search for new particles at higher energies. After at least one false alarm, the team was able to announce a major discovery in June 1977. They had found a new meson with a mass of about 9.5 GeV. His team named it after the Greek letter upsilon (Υ). (Physicists usually distinguish the symbol for the upsilon particle from the letter Y by giving it curly arms.) The identity of the new particle was readily established, since Lederman knew what he was looking for and what its properties were expected to be. By now the existence of quarks was firmly established, and with the tauon pointing the way towards a third generation it was clear that the most likely structure for the new meson was that it was composed of a bottom quark bound to an antibottom quark. This was the first major discovery to be made at Fermilab, just ten years after Wilson had been given the job of building the laboratory from scratch on the Illinois prairie.

The discovery of the upsilon was rapidly confirmed at other laboratories, as they tuned their accelerators to the energy range where the new meson had been found. The interpretation of upsilon as a bottom–antibottom pair was secured when it was shown to be one of a sequence of mesons with similar masses. The charm and anticharm quarks bind together to form charmonium, which can exist in a number of different energy levels corresponding to different mesons. In just the same way, the bottom and antibottom quarks form the bottomonium series of mesons. The bottom quark has a charge of $-\frac{1}{3}$, the same as the strange quark and the down quark, so the bottom quark is the third-generation equivalent of these two lighter quarks. All these types of quark fit naturally into a neat table, as shown in Figure 72 (page 384).[22]

The Dreams of Dr Faustus

> Sitting at my desk or at some café table, I manipulate mathematical expressions and feel like Faust playing with his pentagrams before Mephistopheles arrives. Every once

in a while mathematical abstractions, experimental data, and physical intuitions come together in a definite theory about particles, forces, and symmetries. And every once in an even longer while the theory turns out to be right; sometimes experiments show that nature really does behave the way the theory says it ought.

Steven Weinberg, *Dreams of a Final Theory* (1993)[23]

The discovery of the weak neutral currents at CERN provided strong evidence for the electroweak theory. But the big test would be the existence of the particles that mediate the weak force, the W and Z particles. The job of particle accelerators is to concentrate energy into the tiny regions where particles collide, so that this energy is made available for the production of new particles such as these. However, none of the accelerators under construction in the middle of the 1970s had a hope of attaining the extremely high energies required to produce the heavy W and Z particles. But a determined and ruthless Italian physicist by the name of Carlo Rubbia had a plan to transform one of the world's leading accelerators into a new type of machine that would, he believed, make the production of the W and Z particles a real possibility. Smashing a high-energy beam of particles into a fixed target is a very inefficient process. On impact, only a small proportion of the precious energy that has been concentrated into the beam is made available for the production of new particles. Most of the energy in the beam ends up being transferred into the particle debris that flies onwards on the far side of the target. A much better way to release energy in a particle accelerator is to collide two beams of particles head on. This provides an enormous energy advantage that is even more pronounced when the particles are travelling at close to the speed of light. When two particles slam into each other head on, they are brought to a standstill and release all the energy they have been carrying, which then becomes available for the production of new massive particles. But the technical challenges of a particle collider are also much greater than those of a fixed-target machine. The collider must produce two very intense beams of particles circulating in opposite directions. These beams then have to be directed to intersect, and at the point of intersection they must be focussed extremely tightly so that they don't just pass straight through each other. This would be a formidable feat of engineering.

The idea championed by Rubbia was that a beam of protons could be smashed head on into a beam of antiprotons. The advantage of this configuration is that because anti-protons are negatively charged, their trajectory will bend in the opposite direction to that of a proton in the same magnetic field. A beam of antiprotons would therefore circulate in the beam pipe in the opposite direction to a proton beam, making it possible to convert a proton accelerator into a proton–antiproton collider without the need to construct a second beam pipe with its own set of expensive magnets. Just twenty years since the discovery of the antiproton in 1955, Rubbia was now promoting a plan in which beams of antiprotons could be used to probe the innermost secrets of matter. Many physicists were sceptical about whether the plan was feasible. Trillions of antiprotons would need to be produced, collected and focused into a tight beam no wider than a human hair. But Rubbia and his colleagues believed that they could surmount the technical barriers, and that their collider would have the capability to produce the W and Z particles predicted by the electroweak theory and thereby produce one of the greatest scientific triumphs of the century.

In 1976 Rubbia went to Fermilab to discuss his idea with Bob Wilson, but Wilson decided that he could not afford to divert any of his scarce resources into pursuing the development of antiproton beams. Rubbia then took his proposal to CERN, where a new accelerator, the Super Proton Synchrotron (SPS), had just come into operation. The SPS, built in a seven-kilometre tunnel, was designed to accelerate protons to 400 GeV for use in fixed-target experiments. Pulses of high-energy protons would periodically be directed out of the accelerator ring and along a straight section of beam pipe towards the target area, where their interactions could be studied. Rubbia now presented CERN with an audacious plan for transforming their wonderful new machine into a proton–antiproton collider. In 1979, his persistence paid off and his plan was accepted: the decision was taken to convert the SPS. The upgraded machine would share its time between two modes of operation. It would continue to operate as a fixed-target machine for much of the time, but it would also spend periods operating as a collider.

The reconfiguration of the SPS would also require the construction of new detectors. In a fixed-target machine, most of the energy in the motion of the particles in the beam is transferred to any new particles created as the beam slams into the target material. These new particles then career onwards in the same direction as the beam, so the detector is positioned downstream of the target area to catch the particles, like a wicket-keeper in cricket. By contrast, new particles created in head-on collision events in a collider will scatter outwards from the impact point in all directions, so colliders require large detectors that completely surround the impact point. CERN decided that, if they were to implement Rubbia's ambitious plan, they would need to construct two enormous cylindrical detectors that would analyse the collision products of the SPS. Altogether, around two hundred physicists worked on these detectors. The detectors were known as UA1 and UA2, which stand for Underground Areas 1 and 2. The UA1 detector, which was to be built and operated by Rubbia's team, weighed in at 2,000 tonnes. In an amazing feat of engineering, the detectors were constructed without shutting down the accelerator. Two great caverns were excavated that were large enough to house the detectors well away from the beam line. Although the accurate positioning of the detectors was critical to their operation, they were assembled on trolleys that have been described as 'wheeled chariots', and placed on tracks so that they could be rolled into place when the SPS was operating as a collider, and rolled back again when the SPS reverted to fixed-target mode.

When operating as a collider, the SPS would circulate three bunches of protons in one direction and three bunches of antiprotons in the other direction. They would be accelerated to an energy of 270 GeV before being smashed together head on; but only a fraction of this total energy is released in the collisions. The 270 GeV energy of a proton (or antiproton) would be shared between its constituent quarks and the gluons holding them together. Typically, the energy of each quark might be about 45 GeV. The mutual annihilation of a quark in one of the protons and an antiquark in one of the antiprotons would release about 90 GeV in each collision. If the electroweak theory was correct, this energy would be just sufficient to occasionally create a W or a Z particle, which according to the predictions of the theory should have a mass of around 80 to 90 GeV. There would be thousands of proton–antiproton collisions every second, but this would produce only a handful of W or Z events in a two-month cycle of operation.

By 1981, the SPS was successfully colliding protons and antiprotons, so the search for the W and Z particles could begin. The diagram on the left of Figure 70 shows how a

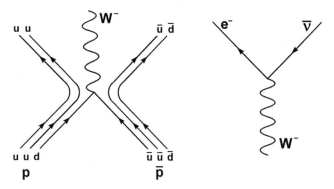

Figure 70 Left: The production of a W-minus in a collision between a proton and an antiproton. Right: The decay of a W-minus into an electron and an antineutrino.

W-minus might be produced in a head-on collision between a proton and an antiproton. The direction of time is upwards in the diagram. The collision involves one of the quark constituents of the proton smashing into one of the antiquark constituents of the antiproton. These two particles will have more or less the same energy, so the head-on collision brings them to a standstill. The W-minus that forms from the energy released by their annihilation will essentially be stationary, existing for a brief period of time before decaying.

There are several ways in which the W particles can decay. All the quarks and leptons feel the weak force, so they all interact with the W and Z particles. Most of the time a W will rapidly decay into a quark and an antiquark that shoot off in opposite directions, although the quark and the antiquark will not be seen directly. The W-minus has an electric charge of −1. As electric charge is conserved in all physical processes, this property can be used to work out the possible quark–antiquark pairings that may be produced when a W-minus decays. The charge of a down quark is −⅓ and the charge of an anti-up quark is −⅔, so this is one possibility: sometimes a W-minus will decay into a down quark and an anti-up quark. Another possibility is that the W-minus will decay into a down quark and an anticharm quark. A third possibility is that it will decay into a strange quark and an anti-up quark.

The Ws usually decay into quarks rather than leptons. But quarks cannot exist as isolated particles, so their independent existence is too brief for them to register in the detector. They are like tiny nuggets around which cascades of mesons and baryons emerge, and these strongly interacting particles are what the detectors see: jets of particles spurting out from the collision point. It is very difficult to disentangle the data produced by these particles. It would be hard to argue with any degree of certainty that they were the result of a decaying W particle. One of the fundamental difficulties in particle physics experiments is to sift out the rare incontestable signal of the particle of interest from the noisy background of debris. As the energy of an accelerator increases, the background particle debris becomes ever more messy and complicated.

Fortunately, the W particles will also occasionally decay in a manner that is much easier to pick out. The W-minus sometimes decays into an electron and an antineutrino. The equivalent decay for a W-plus is into a positron and a neutrino. The electron or positron will stand out from the background of other particles produced in the collision. The diagram on the right in Figure 70 represents the decay of a W-minus into an electron and an antineutrino; one moment the W is there, the next moment it has converted into an electron and an antineutrino. These two particles will each carry away half the energy that was locked up in the mass of the W-minus. The antineutrino will fly off and disappear without a trace, but the electron will be captured by the detector. So the signal that a W-minus has been produced in the collider will be the detection of a very high-energy electron racing away with an energy of around 40 GeV, equal to half the mass of the W-minus. This electron will be conspicuous because its momentum is not balanced by another particle emitted in the opposite direction, as the antineutrino will escape undetected with the other half of the momentum.

Early in 1983, the characteristic decays expected of the W^+ and W^- were identified in both the UA1 and UA2 detectors. The CERN press release heralding the discovery was headed 'Europe forges ahead in particle physics'. It was only a few months later that the Z made its first appearance in both detectors. On the strength of just a handful of indisputable sightings of the heavyweight particles, Rubbia had announced the discovery of the W and the Z particles. But such was the confidence of the teams running the detectors that they were ready to present their results without any hesitation. The Nobel Prize committee was also convinced, and the following year Carlo Rubbia and Simon van der Meer shared the Nobel Prize in Physics. Rubbia's citation was for his efforts in promoting and developing the proton–antiproton collider and his leadership of the UA1 team that discovered the W and Z particles. Van der Meer was recognised for his role in the construction of the antiproton accumulator at CERN,[24] and his invention of stochastic cooling, the technique used to collect and focus the large numbers of antiprotons required to produce an intense beam. In the words of the Nobel committee, they won the award 'for their decisive contributions to the large project, which led to the discovery of the field particles W and Z, communicators of the weak interaction'.[25] This was the first CERN experiment to result in the award of a Nobel Prize.

Building on the success of discovering the W and Z particles, CERN began to construct what would become the biggest particle accelerator in the world, the Large Electron–Positron Collider (LEP). The SPS could produce only a handful of W and Z particles every month, but LEP would be powerful enough to generate them by the cartload. The main purpose of LEP would be to analyse the properties of the W and Z particles and compare them with the precise predictions of the electroweak theory. Any deviations from the expectations of the Standard Model would indicate the need for an improved theory. It was also hoped that LEP might find the Higgs particle, the key to the whole theory. The construction of LEP required the excavation of a new 27-kilometre tunnel straddling the Swiss/French border and four huge underground caverns to house its detectors.

The CERN physicists designed LEP to accelerate a beam of electrons in one direction around the ring and a beam of positrons in the opposite direction. The beams were configured so that they would intersect and produce head-on electron–positron collisions within each of the four detectors. LEP achieved its first collisions in August 1989 and went on to produce copious quantities of W and Z particles. During its decade of operation, its detectors recorded millions of W and Z decays, providing accurate confirmation of the predictions of the Standard Model. Not a single result contradicted the theory. After several upgrades, LEP eventually achieved collision energies of 209 GeV during its last month of operation. Before it was decommissioned, the machine was pushed beyond its design limits in a final attempt to find the Higgs, but without success. It was shut down at the end of 2000 in preparation for the construction of its mighty successor, the Large Hadron Collider.

The Tevatron

PASTORE: Is there anything connected with the hopes of this accelerator that in any way involves the security of this country?

WILSON: No sir, I don't believe so.

PASTORE: Nothing at all?

WILSON: Nothing at all.

PASTORE: It has no value in that respect?

WILSON: It has only to do with the respect with which we regard one another, the dignity of men, our love of culture. It has to do with whether we are good painters, good sculptors, great poets. I mean all the things we really venerate in our country and are patriotic about. It has nothing to do directly with defending the country except to make it worth defending.

Robert R. Wilson's responses to questions from Senator John Pastore on 17 April 1969, in testimony before the Joint Energy Committee of the United States Congress[26]

By 1976, the Main Ring at Fermilab was producing proton beams of 500 GeV, which was the energy that Wilson had been aiming for since the inception of the laboratory. But Wilson was well aware that physicists will always want higher energies. To stay ahead in the race to understand the structure of matter, he had to think about the future. He had been careful when designing the tunnel of the Main Ring to allow sufficient space for later modifications of the accelerator, such as the addition of a second ring of magnets beneath the first. As soon as the Main Ring was functioning to his satisfaction, and he had achieved his goal of 500 GeV beams, Wilson began drawing up plans for the next major energy upgrade. His aim was to use the magic of superconductivity to increase the strength of the magnetic fields in the ring. With superconducting magnets steering the protons around the Main Ring, the energy of the protons could, in theory, be increased to 1,000 GeV. The name for the new machine would be the Doubler.

Superconducting magnets were a very new technology. They held out a great deal of promise for the future, but there were many technological hurdles to overcome before they could be put into operation in such a large project. Although they had been tested in the laboratory, they had never been manufactured on a large scale, and they did not yet have the reliability that was critical if they were to be used in an accelerator. Undaunted, Wilson set up a factory on site where the physicists could design, build and test superconducting magnets very quickly. In the forthcoming years physicists tried out hundreds of different designs at Fermilab until they honed the accelerator magnets to perfection.

Convincing the US Department of Energy to finance the new accelerator proved to be another struggle, even though Wilson stressed the energy-saving features of the design. The annual electricity bill for running the accelerator at Fermilab came to several million dollars, and using superconducting magnets instead of conventional magnets would halve this bill, while enabling the accelerator to produce beams with twice the energy. In 1978 Wilson threatened to resign unless Fermilab could secure sufficient funding for the Doubler. When this funding was not forthcoming he went ahead and offered his resignation, perhaps not expecting that it would be accepted. But it was accepted, and Wilson stepped down. The directorship of Fermilab passed to its leading experimentalist, the co-discoverer of both the muon neutrino and the bottom quark, Leon Lederman. Wilson decided to construct a final sculpture as a gift for Fermilab. To keep the cost down he learned to weld and did most of the construction work himself in his spare time. The end result is the Hyperbolic Obelisk, pictured by the magnificent Wilson Hall in Plate 11.

Despite Wilson's brilliance, Fermilab almost certainly benefited from a change in leadership at this point. Wilson had the vision to create Fermilab. His passionate and charismatic leadership was vital in establishing it as one of the leading particle physics laboratories in the United States, but the challenges of the future, and the funding requirements for larger-scale experiments involving hundreds of researchers and ever-increasing computer processing power, called for a new style and a more diplomatic approach. Lederman proved ideally suited to the task. Under his leadership, Fermilab became the home of the undisputed highest-energy particle collider in the world. Throughout the Wilson, era researchers at Fermilab had to contend with spartan conditions, working for months at a time in waterlogged trenches and living in temporary accommodation. As an experimenter, Lederman had suffered alongside his colleagues and knew the hardships that the researchers had to contend with. He set about improving the working conditions to a standard that would be expected at a world leading research centre.

The Doubler project continued, and Wilson remained as project leader. Eventually Fermilab secured the funding for the accelerator upgrade, and implementation of the new design began in earnest. When it came into operation in 1983, to the great surprise of many people the revolutionary technology in the new accelerator worked like a dream from day one. With its superconducting magnets installed, physicists at Fermilab could now explore particle physics in the TeV realm for the first time. TeV stands for teraelectronvolts – one trillion electronvolts, which is 1,000 GeV or around a thousand times the mass of a proton. The accelerator was renamed the Tevatron. It could race protons and antiprotons around at energies of 1 TeV before smashing them together in head-on collisions. The lead in the particle physics race had been wrested back from Europe.

Few theorists doubted that the third generation of elementary matter particles would eventually be completed, but there was a long delay before its next member was found. It was not until 1995, eighteen years after the discovery of the bottom quark, that Fermilab finally announced the discovery of the top quark. The reason its arrival took so long is that the mass of the top quark is a gargantuan 174 GeV, almost two hundred times the mass of a proton and over forty times the mass of bottom, the next heaviest quark. It had taken two further decades of developments in collider technology to achieve the energies required to produce these behemoths. Because of its enormous mass, the top quark is fantastically unstable. It decays in a mere 0.5×10^{-24} seconds (half a yoctosecond – half of a trillionth of a trillionth of a second), usually into a W-plus and a bottom quark. This is so fast that there is no time for top–antitop mesons to form and therefore, unlike in the cases of charm and bottom, there is no 'toponium' series of mesons. But this fleeting glimpse of the top quark may be the closest we ever get to witnessing a bare isolated quark. Bob Wilson died at the age of eighty-five in January 2000. In a tribute to this giant of twentieth-century physics, the *CERN Courier* described him simply as 'The Magician'.[27] The identification of the fourth and final particle of the third generation, the tauon neutrino, was announced by Fermilab just a few months later.[28]

Does this mean that physicists now expect to find a fourth generation of matter particles? No. There is very strong evidence that there are three and only three generations. LEP measured the parameters of the Standard Model to great accuracy and confirmed the detailed predictions of the

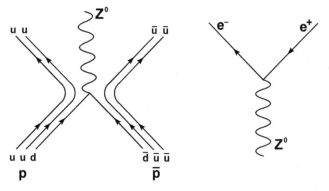

Figure 71 Left: A proton–antiproton collision in which a down quark in the proton annihilates with an antidown quark in the antiproton to produce a Z particle. The remnants of the proton and antiproton are unstable and subsequently decay rapidly into showers of particles that appear as jets in the detector. Right: One of the main decay modes of the Z is into an electron and a positron.

matter particles

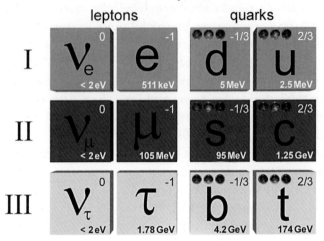

+ antiparticles

force-mediating particles

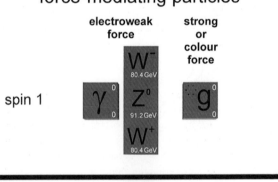

theory. One of its aims was to amass statistics about how the Z particle decays. The Z can decay into a quark and the corresponding antiquark. The quark and antiquark then produce two back-to-back jets of particles. The Z can also decay into a lepton and the corresponding antilepton. For instance, it might decay into an electron and a positron, as shown in Figure 71, or into an electron neutrino and an electron antineutrino. The neutrinos that the Z decays into will not be detected, but the average lifetime of the Z depends on the number of different types of neutrino–antineutrino pair that it can decay into, so by measuring the lifetime of the Z it is possible to deduce this number from experimental observations. One of the great achievements of LEP was to use this method to determine that there are just three types of neutrino and therefore exactly three generations. This assumes that the masses of the neutrinos are much less than the mass of the Z, which is reasonable, since the mass of the Z particle is at least a hundred billion times the mass of the three known types of neutrino. The sensitivity of the Z decays to the number of neutrinos rules out the existence of further generations of elementary matter particles.

Figure 72 (*opposite*) The table of elementary particles in the Standard Model. From left to right: Generation I, the electron neutrino (v_e), electron (e), down quark (d) and up quark (u); Generation II, the muon neutrino (v_μ), muon (μ), strange quark (s) and charm quark (c); and Generation III, the tauon neutrino (v_τ), tauon (τ), bottom quark (b) and top quark (t). Each particle's charge is shown at its upper right and its mass at lower right; mass increases downwards in the table. The force-mediating particles are the photon and the W^+, Z and W^- particles that mediate the electroweak force, and the octet of gluons that mediate the colour force that produces the strong interaction. The final particle in the table is the Higgs (H), which breaks that symmetry of the electroweak force.

The Periodic Table of the Elementary Particles

Dmitri Mendeleyev constructed the original periodic table of the elements in the 1860s. One hundred years later, Murray Gell-Mann organised the subatomic particles that feel the strong force, the mesons and the baryons, into the multiplets of the Eightfold Way. The patterns discerned by Gell-Mann brought order to subatomic physics, just as Mendeleyev had brought order to chemistry. Today, particle physicists can construct a new periodic table of the elementary particles that captures an even deeper level of the structure of matter (Figure 72). This is a good place to summarise these results.

In past classifications, whether the earth, air, fire and water of the ancient Greeks or the chemical elements of Mendeleyev, the fundamental components of matter were considered to be totally distinct to the forces that act on matter. Philosophically, the Standard Model goes much deeper as it places forces on a par with matter. Each force is the result of the exchange of a different type of particle. The table of elementary particles, therefore, contains both the matter particles and force-mediating particles.

There are twelve fundamental matter particles, plus their antiparticles. These particles fall naturally into just three generations of four particles, and that is it! Here is another pattern in the structure of matter, like the patterns in the periodic table and the patterns of the Eightfold Way. But as yet we have no explanation for why there are three generations. It may be a clue to some as yet unsuspected deeper structure to the particles. At the moment no one knows. Perhaps the LHC will enlighten us.[29]

Ignoring gravity, which is too feeble to play a role in particle physics, the Standard Model describes just two forces that act on matter particles. These forces are the electroweak force mediated by the photon and the triplet (W^-, Z^0, W^+); and the strong or colour force mediated by an octet of gluons. The only other fundamental particle is the Higgs, which has the crucial role of breaking the electroweak symmetry and thereby tying the theory together. In the early 1980s, as soon as the triumph of the W and Z discoveries had sunk in, it was already clear what the next really big target for particle physicists would be.

The Superconducting SuperCollider

Around the time that the Tevatron began operation at Fermilab in 1983, American physicists started to put together plans for the next-generation collider. This project would push existing technology as far as it could go with the aim of leapfrogging Europe, where the W and Z particles had recently been discovered. The American plans were extremely ambitious. Their preliminary studies explored the feasibility of constructing a proton collider with a design energy of 20 TeV per beam. The project pushed every superlative to the limit, so the machine was duly named the Superconducting SuperCollider (SSC). By the end of the decade the design for an 87-kilometre oval circuit had been finalised, and in March 1990 the State of Texas began acquiring 16,000 acres of land at the chosen site, south of Dallas at Waxahachie.[30]

One of the main tasks of the SSC was to be the search for the Higgs, the last missing piece of the Standard Model jigsaw. In support of the project Leon Lederman wrote a book with the provocative title *The God Particle*, Lederman's dubious nickname for the Higgs.[31] But it failed to convince the doubters in Washington. By 1993 the projected cost of the SSC had risen to around $12 billion. Around $2 billion had already been spent, and over 23 kilometres of tunnels had been completed when the House of Representatives voted to cancel the project. Meanwhile, in Europe plans were under way for the construction of the Large Hadron Collider at CERN. This incredible project and its quest for the Higgs is the subject of the next chapter.

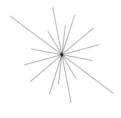

Chapter Nine

THE ULTIMATE PRIZE

In the course of giving a very large acceleration to our
particles, let us hope that we can contribute at least a small
acceleration to society.

Robert R. Wilson, policy statement (1968)[1]

The World Wide Web

If we want to find out the latest news about the Large
Hadron Collider, a good place to start is the CERN website.[2] Of course, when the grand synthesis that is the Standard Model was created in the 1970s, there was no such thing
as the World Wide Web. Even the breakthrough discoveries of the W and Z particles in 1983 arrived well before its

birth. The World Wide Web was first proposed by British computer scientist Tim Berners-Lee while he was working at CERN in 1989, so the CERN website became the first website anywhere in the world. CERN had already become the biggest internet node in Europe, and its physicists were drowning in data. The amount of information they needed to access was increasing rapidly, and Berners-Lee realised that the internet offered the perfect tool for managing this information. What was required was software that was free and standardised that would enable people to share information efficiently across all the different types of computer that were being connected together.

The resources that CERN allocated to the realisation of Tim Berners-Lee's vision were very limited, but despite this the World Wide Web was a phenomenal and almost overnight success. It was clearly an idea whose time had come. But, had the Web not evolved within an organisation that was dedicated to pure research, and as the brainchild of such a public-spirited individual, it might well have turned into a very different entity. The other networks that we rely on so much, such as our electricity grid, radio and television networks, and telephone systems are all centrally controlled. With a large government or international corporation in control, the information superhighway would not have become today's vast open network, effectively free to everyone for whatever purpose, commercial or otherwise.

By 1996, just seven years after Berners-Lee first proposed the Web, the Director General of CERN, Chris Llewellyn-Smith, calculated that the value of the Web exceeded the total contributions to CERN of all the member states in the four decades since its inception in the 1950s.[3] The Web has continued its incredible growth and has now become a vital part of all our lives. The impact of Tim Berners-Lee's invention is put into context when we consider that the full construction and operating costs of the Large Hadron Collider, shared by the twenty member states, are estimated to be around €5 billion. Today there are several web businesses that have become household names: Google, Amazon, eBay, Facebook and Yahoo. Each has a stock market valuation that exceeds the cost of the LHC by a very wide margin. Google alone is valued at well over €100 billion.

The World's Greatest Colliderscope

In the early 1980s, while the Large Electron–Positron Collider (LEP) was still under construction, there were groups at CERN who were looking even further into the future of particle physics. In an attempt to stay in the race with the United States, where the Superconducting Super Collider was being planned, CERN physicists designed the Large Hadron Collider. The LHC would take over the 27-kilometre LEP tunnel when LEP reached the end of its eleven-year lifespan. During the early period of the explosive growth of the Web, in 1993 and 1994, the future of CERN hung in the balance. If the member states agreed to fund the next-generation collider, the laboratory would have a secure future at the forefront of particle physics for many decades to come. When the United States cancelled the Superconducting Super Collider in 1993, the European collider assumed an even greater global significance. If the European states took a similar view about the LHC, then for the foreseeable future particle physics would be in the doldrums throughout the world. After strenuous negotiations, Chris Llewellyn-Smith announced in December 1994 that, thankfully,

CERN's governing body had taken a unanimous decision to proceed with the construction of the LHC, with a date of 2005 pencilled in for the start-up of the machine. The main goal of the new collider would be the discovery of the Higgs particle, the vital component of the electroweak theory that was responsible for breaking the symmetry between the electromagnetic and weak forces.

In September 2000, LEP was due to close down and make way for the LHC. The electron–positron collider had been searching for signs of the Higgs, by now the last missing particle in the Standard Model's table of elementary particles. During its final term of operation, physicists pulled out all the stops to run the accelerator at well beyond its original design energy. LEP had ruled out a mass for the Higgs of less than 100 GeV, but there were tantalising hints that its mass might not be much greater than this. Some physicists believed that with just a little more time LEP might find the Higgs. There was a debate within CERN about whether the life of LEP should be extended into 2001 with the aim of securing this momentous discovery. In the end, LEP was given just six extra weeks of operation, and the decision was taken that it would be decommissioned at the end of 2000 so that the construction of the LHC could begin. It was a big decision. In 1995 the Tevatron at Fermilab had tracked down the top quark, with its whopping mass of 174 GeV, and had followed this success with an announcement of the discovery of the tauon neutrino in July 2000, just months before LEP was scheduled to be switched off. The race was now on. Could the Europeans build the LHC before the Americans triumphed again by tracking down the Higgs particle?

The Large Hadron Collider is the biggest, boldest and most complex scientific enterprise ever attempted – and by

2001 the enormous project was already behind schedule. CERN decided to cut back on all non-LHC activity and delay the LHC's start date until 2007, which was still a very tight schedule for the installation of such new and complicated technology. In 2007, problems with some of the focusing magnets forced the switch-on date back even further. Finally, a date of 10 September 2008 was set for the launch of the greatest particle accelerator the world had ever seen. The day arrived, and, amid much excitement and wild speculation throughout the media, the first proton beams circulated successfully in the ring of the LHC. But, just over a week later, on 19 September and with the world still watching, one of the superconducting magnets failed catastrophically as its temperature rose above its critical temperature. The energy stored in its magnetic field was released in an instant, rapidly heating up the magnet. Tonnes of liquid helium coolant boiled and leaked explosively into the LHC tunnel, blasting the magnet from its mount.

Initially, the repair work was expected to take a couple of months, as it required warming an entire three-kilometre sector of the cryogenic system to room temperature. But far more extensive repairs were soon found to be necessary. The cause of the magnet failure was tracked down to an electrical fault caused by lack of solder on the superconducting wires that connect the magnets. Further investigation revealed many more such soldering faults. The resumption of LHC operations was put back to November 2009, but when the mighty machine was restarted it worked like a dream. The first particle collisions took place one month later. The machine was shut down briefly for maintenance during the early months of 2010, and then in April the LHC began its work in earnest.

The LHC – A Monster Machine

> One Ring to rule them all, One Ring to find them,
> One Ring to bring them all and in the darkness bind them.

J.R.R. Tolkien, *The Fellowship of the Ring* (1954)[4]

The simplest way to accelerate charged particles is in a straight line in a machine which is known, quite naturally, as a linear accelerator. These machines, first constructed in the 1930s, consist of a series of electrically charged metal plates with holes in their centres through which beams of charged particles are accelerated. Protons are positively charged, so a bunch of protons will be repelled from a positively charged plate and attracted to a negatively charged plate. As the protons are accelerated towards the negatively charged plate their energy will increase. The key to the operation of the linear accelerator is that the electrical polarity of the plates is switched as the proton bunch passes through the hole in their centre. The bunch will then be repelled from the plate through which it is passing and attracted to the next plate along its trajectory. When it reaches this next plate, the polarities of the plates are switched again and the bunch of protons is boosted towards the following plate. By timing the flipping of the polarity to coincide with the passage of the proton bunches, energy is transferred to the particles and they are accelerated to higher energies.

In a linear accelerator, charged particles must be accelerated from a standstill on each run before being smashed into their target. A more efficient design is to bend the accelerator into a circle and guide the particles into a closed loop, so that the acceleration can be continued for many circuits around the machine. Applying a magnetic field in just the right way, will curve their path of the particles so that it closes on itself and forms a circle. As the energy of the particles in the accelerator ring increases, the magnetic field must be increased in step so that there is just the right force on the particles to keep them circulating within the beam pipe. Because the increasing strength of the magnetic field must be synchronised with the acceleration of the particles, these machines are known as synchrotrons. Keeping the particles in a ring has the advantage that their acceleration can take place in stages over many circuits. Also, once the maximum energy has been achieved, the circulation of the beams can be maintained for extensive periods of time.

A bigger circuit has a more gradual curvature, so lower magnetic fields are required to bend the path of the charged particles around the accelerator. Once the size of the circuit has been fixed, it is the strength of the magnets that limits the maximum energy that can be attained by the particles. If the particles were accelerated to energies beyond this limit, the magnets would not be strong enough to bend their path sufficiently to keep them within the accelerator ring. If we want to produce the most powerful accelerator possible, we should select the largest ring and surround it with the most powerful magnets that our purse will allow. If the magnets are superconducting, this will have the added benefit of dramatically reducing energy losses due to electrical resistance.

To keep construction costs down, the LHC is reusing the tunnel that was originally excavated for the LEP collider. It lies between 50 metres and 175 metres below ground and straddles the French/Swiss border to the north-west of Geneva. The tunnel is just under 27 kilometres long and nearly four metres in diameter, which makes it comparable in size to the Circle Line of the London Underground. Within the tunnel is a stainless steel pipeline with a diameter of just under one metre. The pipeline contains the entire apparatus of the LHC: the proton beam pipes, the bending and focusing magnets, the cryogenic system for cooling these superconducting magnets, and the electronics for controlling each component.

The LHC is the final link in a chain of accelerators (see Figure 73). Pulses of protons pass through a complex of four

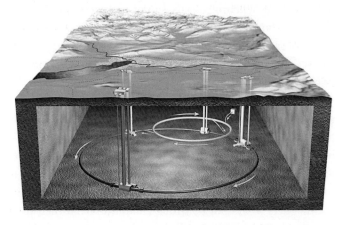

Figure 73 Schematic layout of the LHC complex at CERN. The arrows represent the direction travelled by the protons, which enter the Super Proton Synchrotron (SPS, shown as the smaller circle) from the Proton Synchrotron (not shown). The protons are then injected into the Large Hadron Collider (LHC, shown as the large circle) in both directions. The sites of the four detectors are also shown: ATLAS is within the circle of the SPS; clockwise from there are ALICE, CMS and LHCb.

older machines before forming the two counter-rotating beams of the LHC. The protons start off as the nuclei of the atoms in hydrogen gas. They are stripped of their electrons in a high electric field, and are then fed into a linear accelerator, designated Linac 2, and accelerated to 50 MeV, at which energy they are injected into the Booster. They circulate in the Booster until they reach 1.4 GeV. The protons then pass on to CERN's oldest accelerator, the Proton Synchrotron, which takes them up to 25 GeV. They are then guided into the Super Proton Synchrotron (SPS), built in the 1970s. When the SPS has accelerated the protons up to an energy of 450 GeV, they are injected into the two beam pipes of the Large Hadron Collider. The two beams are not continuous. Once the LHC is operating at full tilt they will consist of up to 2,800 simultaneously circulating bunches of protons. Each bunch will be a couple of centimetres long and contain roughly a hundred billion protons. The final acceleration of the protons will then begin. As the energy of the protons increases, the magnetic field produced by the bending magnets must increase in step, until, after twenty minutes of acceleration in the LHC, the beams will attain the energy of 7 TeV,[5] an enormous energy that is deserving of the prefix T for 'tera-', from the Greek word for monster. The bending magnets will now have reached their maximum field strength, which is about 170,000 times the strength of the Earth's magnetic field. This magnetic field is just sufficient to steer protons with an energy of 7 TeV around the LHC ring, and this is what determines the design energy of the LHC and makes it the most powerful particle accelerator the world has ever seen. In its first period of operation, during 2010 and 2011, the LHC accelerated protons halfway to this energy, producing collisions at an energy of 3.5 TeV per beam. For the 2012 run the beam energy was raised to 4 TeV.

Operating the LHC requires a prodigious energy supply. Its electricity consumption is around 120 MW, out of a total of 230 MW for the whole CERN laboratory. This is equal to the consumption of the whole of the rest of the Swiss canton of Geneva with its population of almost half a million. For this reason, the original plan was to shut down the LHC during the winter months when electricity supplies are most expensive. This idea has now been abandoned in order to optimise the use of the accelerator. A large proportion of the energy is consumed by the cryogenic system that keeps the superconducting magnets at an ultra-low temperature, but overall the use of this almost miraculous superconducting technology offers dramatic savings in energy. The result is that the energy consumption of the LHC is not much greater than that of the Super Proton Synchrotron, even though the LHC accelerates protons to much higher energies.

The total amount of energy concentrated into each of the proton beams in the LHC is comparable to that carried by a British high-speed train travelling at 200 kilometres per hour. (Even this is dwarfed by the energy stored in the magnets of the LHC, which is around thirty times as much – about the same as the energy of motion of the world's largest supertanker when fully laden, or a large US aircraft carrier cruising at its top speed of around 50 kilometres per hour.) Amazingly, all this energy is carried by a miniscule amount of matter. There will be several hundred trillion protons circulating in the LHC at any one time. An enormous number – but protons are tiny. In a year of operation, the total mass of protons accelerated around the LHC will amount to just a microgram, an amount of matter that would easily fit on a pinhead. It is about the mass of the ink in a full stop.

Our understanding of the equivalence of mass and energy dates back to the revolution in physics initiated by Einstein a hundred years ago. As highlighted in Chapter 5, the fusion of hydrogen nuclei into helium nuclei produces the vast energy output of the Sun. In the process of this conversion a small percentage of the mass is released as energy. We quite rightly consider even a small amount of mass to be equivalent to a huge amount of energy. This puts into context the degree to which the LHC concentrates energy into its proton beams. The mass of a proton is just under 1 GeV. Within the LHC the protons will each be carrying an energy of 7 TeV, or 7,000 GeV. The total energy carried by the protons is therefore over seven thousand times the energy that is stored in their mass.

Massive particles, such as protons, can never attain the speed of light, but as their energy increases in an accelerator, they edge ever closer towards light speed. At the top energy of the LHC, the protons will be racing around the circuit within a whisker of this cosmic speed limit. Their speed will be something like 99.999 991 per cent of the speed of light. At this tremendous speed, they will complete over 11,000 revolutions of the accelerator every second. Within the beam pipe a vacuum must be maintained that is emptier than inter-galactic space. Otherwise the protons would be scattered by collisions with the gas molecules in the pipe.

The factor that limits the energy that can be achieved in a proton accelerator ring is the strength of the magnets that are used to bend the trajectory of the protons. To achieve the greatest possible acceleration at the LHC, the entire circuit is surrounded by superconducting magnets, utilis-ing the technology pioneered in the Tevatron at Fermilab. In fact, most of the superconducting magnets at the LHC were manufactured at Fermilab. The LHC contains several thousand superconducting magnets which guide and focus the proton beams. The largest of these magnets are the 1,232 bending magnets, made of an alloy of niobium and titanium. The total length of the superconducting cables within them is a continent-spanning 7,600 kilometres. To achieve their design magnetic field, the superconducting magnets must be refrigerated to a temperature of 1.9 K, just 1.9 degrees above absolute zero. Twelve million litres of liquid nitrogen must be evaporated to cool the 38,000 tonnes of machinery that comprises the LHC to within two degrees of absolute zero in the initial cool-down. The cryogenic system that maintains the magnets at this extremely low temperature is divided into eight octants, each of which is over three kilometres long and larger than any such system previously constructed. Its integrity depends on tens of thousands of leak-proof pipe junctions. The engineering challenge of maintaining cryogenic temperatures is compounded by the fact that the machine is bathed in radiation emitted by the circulating proton beams. All the heat generated by this radiation must be removed rapidly to ensure that the magnets remain ultra-cold. Around 700,000 litres of superfluid liquid helium circulates through the cryogenic system to dissipate this heat.

The bending magnets themselves are ingeniously designed to contain two separate magnetic channels within one physical structure, with the magnetic fields in these channels running in opposite directions. In this elegant arrangement, the force curving the path of the proton beam in one channel is in the same direction as the force curving the path of the counter-rotating proton beam in the other channel. Therefore both beams of protons can be steered around the ring with just one set of magnets (see Figure 74).

At four sites around the ring, the two proton beams are focused extremely sharply and guided together so that they cross. At these intersection points, the hundred billion protons in a bunch are squeezed down into a beam with a width of just 16 micrometres – much finer than the finest human hair. Even so, protons are so tiny that the vast majority of the

Figure 74 The inside of the LHC tunnel during construction and before the full circuit was completed. The stainless steel pipeline contains the beam pipes, the superconducting magnets, the cryogenic system and the control systems of the LHC. The blocked-off ends of the two beam pipes can be seen side by side emerging from the pipeline.

protons in the two beams whizz past one another without colliding and continue to circulate in the beam pipe. Each time the paths of two bunches intersect, there are typically just twenty or so proton–proton collisions.

There are two properties that characterise the performance of a particle collider. One is the energy at which collisions take place. The other is the luminosity of the collider, which is the rate at which collisions are generated within the detectors – the intensity of the blaze raging in the heart of the machine. The luminosity of the LHC will be ramped up gradually as the machine is fine-tuned. This means that one month of operation in 2011 may produce the same volume of data as the whole of 2010, and by the end of 2012, a single month may produce as much data as the whole of 2011. When the LHC is operating at full capacity, proton bunches will cross every 25 nanoseconds, so there will be 40 million bunch crossings every second, producing almost a billion collision events per second within each detector. The proton beams will circulate for up to ten hours, making over four hundred million revolutions of the machine, during which time the volume of collisions will be vast. Analysis of the particle debris produced in the collisions begins in the detectors that have been built at each of the four intersection sites around the LHC circuit. These detectors are larger than any previously constructed at an accelerator facility. Each has required the excavation of its own subterranean chasm.

ATLAS – A Giant Particle Detector

Here Atlas reign'd, of more than human size,
And in his kingdom the world's limit lies.

Ovid, *Metamorphoses*[6]

The four main detectors at the LHC have different designs so that they can analyse the physics of the collisions in complementary ways. ATLAS and CMS (Compact Muon Solenoid) are general-purpose detectors. The other two, LHCb (the 'b' stands for bottom, as in bottom quark, which is the main target of the detector) and ALICE (A Large Ion Collider Experiment) are designed to study specific types of physics. LHCb will concentrate on the physics of the bottom quark, which will show up in great profusion in the proton–proton collisions at the LHC. These studies are designed to shed light on the mystery of why the universe is built entirely out of matter. This observed matter–antimatter asymmetry is still not fully understood. The LHC will not only be used as a proton collider: it will also be used to smash together beams of extremely high-energy nuclei from atoms of lead. ALICE is designed to investigate the dramatic results of these heavy-ion impacts. The first use of the LHC in this mode took place late in 2010. The experiments are designed to explore a new form of matter that exists at extremely high temperatures, known as a quark–gluon plasma, in which quarks and gluons are emancipated from their hadron cages and exist briefly as independent particles.

ATLAS is the namesake of the primordial titan who supports the heavens on his shoulders, and as such it is fitting that it is the biggest of the detectors at the LHC, weighing in at around 7,000 tonnes – which is as much as ten fully laden Airbus A380 'superjumbo' aircraft. This huge barrel-shaped instrument has the largest volume of any detector ever built at a particle collider. It is 46 metres long and 25 metres in diameter (see Plate 12 and Figure 75). The name of this gargantuan device comes from the rather contrived acronym 'A Toroidal LHC ApparatuS'. Almost three thousand researchers from thirty-five countries are involved in the ATLAS collaboration.[7]

ATLAS is designed to detect all the highly energetic particles that emerge from the proton–proton collisions and to measure as precisely as possible their trajectories in three dimensions. The aim is that no detectable particle should be missed. This is a tall order, as two proton bunches cross within ATLAS every 25 nanoseconds, producing a set of around twenty extremely high-energy proton–proton collisions, each of which results in an explosive splurge of high-energy particles racing away from the impact point in all directions. The result is a pyrotechnic display within the detector of around a thousand newly created particles.

The electronics in the detectors must be extremely fast to cope with the rate at which the collisions are being generated. To provide accurate information about the locations at which the particles are produced and then decay, the detectors must have a very high spatial resolution. They must also be extremely resistant to radiation damage so as to withstand the relentless high-energy bombardment they will receive for many years. It was a major challenge to ensure that all the components of the detectors would continue to function under prolonged exposure to intense radiation, and this determined which materials could be used in their manufacture. An unsuitable material would be one whose atomic nuclei would readily transform into different nuclei when bathed in a barrage of protons and neutrons or assaulted by other sorts of particles, as this would change them into different types of atom with different physical properties.

ATLAS must determine the energy and the identity of each of the particles that are generated in the proton–proton collisions. It sits in a magnetic field that curves the trajectories of the charged particles that pass through it. This provides information about the electric charge of the particles and their energy. The paths of very high-energy particles are almost straight, while lower-energy particles curve into tight spirals. The particle showers produced in the LHC are so energetic that a very strong magnetic field is required to produce any significant curvature of the particle paths, so ATLAS is embedded within its own set of eight 25-metre-long superconducting magnets that form a cylindrical magnetic field around the beam pipe at the heart of the detector.[8] These powerful magnets enable the machine to analyse and catalogue the particle debris as it flies through the detector.

Matryoshka Giants

ATLAS is built up in layers, each designed to recognise different types of particles. There are four nested subsystems arranged concentrically like Russian matryoshka dolls: the inner tracking detector, the electromagnetic calorimeter, the hadron calorimeter and the muon chamber, as shown in Figure 75.

Interpreting particle interactions requires very accurate information about particle positions and trajectories in

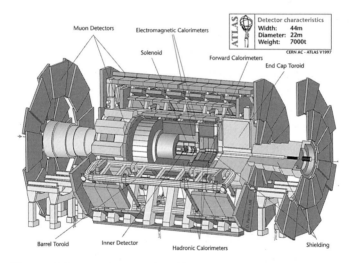

Figure 75 Cutaway diagram of the ATLAS detector. The tubes labelled 'Barrel Toroid' in the diagram contain the superconducting magnets and can be seen clearly in the photograph of ATLAS in Figure 74. The beam pipes enter the diagram from the left and the right, and the two beams intersect at the centre of the detector.

the innermost region of the detector, close to the collision point. It is the job of the tracking detector (labelled 'Inner Detector' in Figure 75) to accurately determine the trajectory of all the charged particles emanating from the collisions within the detector. This part of the detector is like an enormous high-resolution digital camera. It is constructed from the most expensive and highest-resolution components: arrays of silicon detectors very similar to the chips used in digital cameras with a high pixel density. The chip in a digital camera detects photons as they fall on the chip and records the point at which each photon hits the chip. When a photograph is taken, the chip records the number of photons that arrive at each pixel – which is just a cell in a square array on the chip – during the exposure time of the photograph. This information is converted into the brightness of each pixel in the photograph. (Taking a colour photograph requires the use of filters, each of which allows photons in a different frequency range, or colour, to impinge on the chip.) The silicon detectors in ATLAS consist of layer upon layer of these chips. The trajectory of a charged particle is recorded as it passes through a pixel in each layer. When the information from all the layers is collected, it provides a high-resolution track of the particle's passage through the three-dimensional space within the detector. These paths will not be directly visible, but will be reconstructed electronically and displayed on computer monitors.

The next layer of the detector, the electromagnetic calorimeter, measures the energy of particles that interact via the electromagnetic force. It detects particles such as photons, electrons, positrons, protons and other charged particles. In this region the brakes are applied to each particle, bringing it to rest in order to determine how much energy it was carrying. This part of the detector consists of layers of lead interleaved with layers of liquid argon. The passage of a particle through the lead generates showers of secondary charged particles. These particles then stream through the liquid argon, knocking electrons out of the argon atoms. The flow of the resulting argon ions produces an electrical current that is measured in the surrounding metal plates. The strength of this current indicates the energy of the original particle that entered the calorimeter.

The hadron calorimeter works along similar principles to the electromagnetic calorimeter, but it is designed to measure the particles that interact via the strong force. For this reason it is constructed from iron – a material whose dense atomic nuclei will slow down, but not absorb, any incident hadrons, such as protons, antiprotons, neutrons and pions. These particles are brought to a halt in the hadron calorimeter so that their total energy can be determined. Plastic scintillator tiles embedded within the iron absorber take this measurement. These are the modern equivalent of the zinc sulphide scintillators used by Rutherford and his colleagues to detect the scattered alpha particles in their gold foil experiments. Each time a particle hits the scintillator, it releases a brief pulse of light. It was viewing such flashes that enabled Geiger and Marsden to see that some of the alpha particles were bouncing back from the gold foil. This was the clue that enabled Rutherford to deduce the existence of a dense nucleus within the heart of the atom. Today, the flashes are channelled down optical fibres and counted electronically.

Muons are weakly interacting, which makes them highly penetrating. They will pass through both the electromagnetic and the hadron calorimeters. Muons are the only particles other than neutrinos that are able to make their way through the three inner layers of the detector without decaying or being completely absorbed. The outer layer of ATLAS is the muon chamber, which has been specifically designed to track the muons. The information it gathers about their trajectories is vitally important for the detection of the Higgs particle.

Each particle leaves its own distinctive signature in the ATLAS detector, as indicated in Figure 76. Photons are not charged, so they do not show up in the tracking chamber,

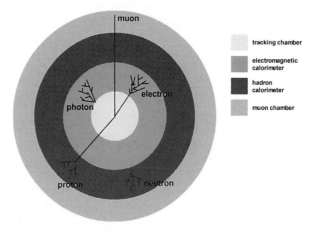

Figure 76 A schematic diagram of the signatures of some of the different types of particle detected in the four layers of the ATLAS detector. In reality the paths are curved by the magnetic field within the detector, and the direction of curvature determines whether they have been produced by positively or negatively charged particles.

but they do produce showers in the electromagnetic calorimeter. The paths followed by electrons and positrons curve in opposite directions in the magnetic field in the detector; their tracks can be seen in the tracking chamber. They also produce showers in the electromagnetic calorimeter.

Neutrons are electrically neutral. They do not produce tracks in the tracking chamber and they do not produce showers in the electromagnetic calorimeter, but they do interact via the strong force, so they show up as showers in the hadron calorimeter. Protons can be seen in the tracking chamber, the electromagnetic calorimeter and the hadron calorimeter. The elusive neutrinos disappear without a trace and do not interact with any of the components of the detector. For this reason it is important that in each collision event ATLAS detects every particle that is detectable, which means all known particles except for the three types of neutrino. If everything else is tracked down, then, when particle decays are reconstructed, at any point where energy and momentum seem to have disappeared, it can be concluded that a neutrino has sped off with some of the energy. This assumption is reliable only if there is confidence that everything other than neutrinos has been detected.

In the 25 nanoseconds between proton bunch crossings in the LHC, light travels just 7.5 metres, which is much less than the size of ATLAS. So even though the particle debris being produced in the LHC is travelling at virtually the speed of light, the particles generated in a second set of collisions will enter the detector before the particles produced in the previous set have exited the detector.[9] Clearly, this presents an enormous challenge for real-time analysis of the data. The functioning of the thousands of elements that form the huge ATLAS detector must be synchronised to within a single nanosecond (one-billionth of a second). Immediately after each collision, as the particle debris impinges on the detector, every interaction of a particle with the detector is time-stamped and fed into a buffer, where it is held and analysed electronically. From these scattered fragments of data, collected and recorded throughout the detector, ATLAS determines the trajectory of each particle and electronically reconstructs the details of the collision event.

The Quest for the Higgs

The Standard Model encapsulates all our current knowledge of particle physics. It represents the best information we have about the fundamental structure of matter. It combines theories of the colour force and the electroweak force with the three generations of matter particles that these forces act upon. The colour force is described by quantum chromodynamics. It is the force that holds quarks together to form protons and neutrons, and it is also the force that holds the protons and neutrons together to form the nuclei of atoms. The electroweak force is described by the Glashow–Weinberg–Salam theory, which offers a unified understanding of electromagnetism and the weak force. The electromagnetic force is all-pervasive in our lives. We invoke its power whenever we lift a finger and whenever we use that finger to switch on a light. The weak force plays a vital role in the transmutation of the elements within stars and their dispersal throughout our galaxy. Without the weak force, there would have been no complex chemical soup from which life could arise on our planet. But the symmetry between the electromagnetic and weak forces is broken, and, according to the Standard Model, the existence of the Higgs particle is required to perform the symmetry breaking.

The Standard Model table of elementary particles consists of three generations of matter particles. In the first we find the up and down quarks, the electron and the electron neutrino. Then there are two more generations of particles that appear to be two sets of heavier replicas of these. We also have a collection of particles that produce the forces. The colour force is mediated by an octet of gluons, and the electroweak force is mediated by the photon and the W-minus, W-plus and Z-nought particles. The Higgs occupies the final slot in the table.

The Signature of the Higgs

> We have this feeling we are right at the threshold of some great discovery. It is like Columbus, if we get to the beach it will be very exciting. We won't be able to explore the full continent behind it, that is the job of the LHC.

Piermaria Oddone, Director of Fermilab (2009)[10]

What happens when two extremely high-energy protons collide? When the LHC is operating at full power, the protons in the two beams will be accelerated to an energy of 7 TeV. In the head-on impact of two such protons, a total energy of 14 TeV is released. However, protons are composite particles, formed of a collection of quarks held together by a tangle of gluons. About half the energy of a proton is carried by the gluons and the rest is shared between the quarks. In a collision between two protons, only one constituent particle in each proton will take part in the collision. So the events will actually be quark–quark, quark–gluon or gluon–gluon collisions. Typically, the quark or gluon components of the protons that undergo collisions at the LHC will each be carrying about 1 TeV of energy, so around 2 TeV will be released in their head-on collisions.

The two high-energy proton fragments that were not directly involved in the collision are now extremely unstable. They rapidly decay via the colour force, producing jets of particles that spray out from the impact point. A large proportion of the debris produced in the proton–proton collisions results from the collateral damage from the decay of the proton fragments, not from the actual quark–quark, quark–gluon or gluon–gluon collision that physicists are really interested in. This makes the collisions very messy. By comparison, collisions at the LHC's predecessor, LEP, were very clean. LEP was an electron–positron collider. Electrons and their antiparticles positrons have no substructure, so each impact was a straightforward interaction between an electron and a positron in which all the energy carried by the colliding particles was released in the collision. Interpretation of the data from the LHC will be vastly more difficult than it was for LEP. Each collision is expected to result in around fifty particle trajectories being recorded in the detector.

The Standard Model offered quite precise predictions about the mass of the W and Z particles, which was a great help in tracking them down. It is unlikely that Carlo Rubbia would have persuaded CERN to go ahead with the re-engineering of their SPS accelerator without a clear indication of the mass range within which these particles would be discovered. By contrast, the Standard Model does not pin down the mass of the Higgs. LEP showed that the mass of the Higgs could not be less than 114 GeV, but it was possible that the Higgs was just out of the reach of LEP, with a mass not much greater than this. At the other extreme, Standard Model calculations ruled out a mass for the Higgs greater than 600 GeV. But this still left an enormous range of energies for the LHC to explore, within which the Higgs might appear anywhere.

By 2011, results from the Tevatron had ruled out the existence of the Higgs with a mass between around 150 and 180 GeV, but it was clear that with the LHC working so well, and amassing mountains of data, the Tevatron could no longer compete. On 30 September 2011 it produced its final proton collisions and decommissioning began. This left the LHC without a competitor in the race for the Higgs, and there was still a huge energy range in which to search.

According to the Standard Model, the most common way to produce the Higgs in the LHC is not through the collision of two quarks, but through the collision of two gluons.[11] This process is depicted by the Feynman diagram in Figure 77. Here the two gluons, represented as curly lines, collide head

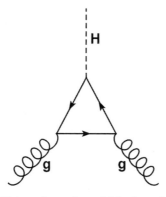

Figure 77 The collision of two gluons (g) leads to the production of a Higgs (H). The triangular loop could be formed by quarks or antiquarks of any flavour, but the top quark loop gives the biggest contribution. This process is known as gluon fusion.

on. The outcome of the gluon collision is the emission of a Higgs, which is shown as the dashed line labelled H. Gluons do not interact via the electroweak force, so the loop of quarks is required to couple them to the Higgs.

But the Higgs itself will not register in ATLAS or the other detectors, so to track it down it is necessary to consider how it decays. The search concentrates on the decay modes that produce particles that will stand out in the detectors.

The Standard Model encapsulates all the information about how the Higgs is expected to behave. It must interact with all the particles that feel the electroweak force, which implies that there are many ways in which it can decay. However, the Higgs cannot decay into two particles if the total mass of these two particles is greater than its own mass, so how the Higgs decays depends critically on its mass. The Standard Model predictions for Higgs decays have been built into computer simulations that can be compared with the actual data now being generated by the LHC.

One of the characteristic properties of the Higgs is that it is responsible for conferring mass on all elementary particles.[12] According to the electroweak theory, the mass of all the fundamental particles is produced by the pull of the Higgs force as they continually stream through the background Higgs field. And the 'stickier' the particle finds the Higgs field, the greater will be its mass. This means that heavier particles must have much stronger interactions with the Higgs. For instance, the mass of the muon is about 207 times the mass of an electron, so the muon is expected to interact with the Higgs about 207 times as strongly as an electron does. This information tells us how we can expect the Higgs to decay most often. Quarks are heavier than leptons, so there will be many more decays into a quark–antiquark pair than into a lepton–antilepton pair, and the most common types of quark to be found in the Higgs decays will be the heaviest ones.

The vast majority of Higgs decays will be into a bottom quark and an antibottom quark, simply because these are the heaviest particles it can decay into. This decay mode gives rise to two complicated back-to-back jets formed of numerous particles. Unfortunately, this is not much help in identifying the Higgs. The detectors at the LHC are swamped with bottom quarks as there are many ways in which bottom quarks can be produced in proton–proton collisions. It is impossible to distinguish the bottom quarks produced in Higgs decays from those produced in other ways.[13] Although the bottom quark was discovered as recently as 1977, in the LHC it will be produced in such profusion that it forms the messy background that physicists (or at least their electronics) must sift through in order to find interesting new physics. The great discovery of one generation has become the annoying background of the next.

To spot the Higgs, CERN's researchers need a clean decay signature. Strong interactions are Byzantine in their intricacy. They produce a plethora of particles whose identity is difficult to disentangle. In order to avoid these complications, physicists will be focusing on decays that do not involve quarks and gluons – the particles that feel the strong force. The clearest signal will come from a decay into four leptons via a pair of Z particle intermediaries. This will be far from the most common way in which the Higgs will decay, but it will stand out. The total energy of the four leptons will equal the mass of the Higgs. Figure 78 shows some examples of this type of

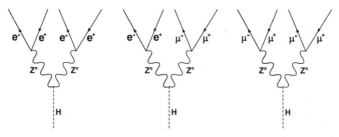

Figure 78 Three examples of how the Higgs particle can decay into two pairs of leptons. Left: The Higgs decays into two electron–positron pairs. Centre: The Higgs decays into an electron–positron pair and a muon–antimuon pair. Right: The Higgs decays into two muon–antimuon pairs.

decay. Each Z could give rise to an electron and a positron, a muon and an antimuon, a tauon and an antitauon, or any of the three different types of neutrino and antineutrino.

Detecting a high-energy muon–antimuon pair or an electron–positron pair is easy in ATLAS or CMS, and this provides a clear signal that can be identified as the Higgs. A simulation of such a Higgs decay within ATLAS is shown in the mural in Plate 13. Of course, finding one example of such an event does not clinch the argument for the discovery of the Higgs. Numerous examples of these lepton quartets must be seen, and if the sum of their energies always equals the same value, it would be safe to conclude that the Higgs has been found and that this total energy is the mass of the Higgs. These events are clear, but they are also much rarer than Higgs decays into heavy quarks, so tracking down the Higgs would require a couple of years of data gathering and statistical analysis. The LHC is producing vast amounts of data which all needs to be disentangled. This is one of the biggest challenges of the project.

At the July 2011 conference of the European Physical Society, researchers created a flurry of excitement by suggesting that the first 'tantalising' hints of the Higgs particle might have been picked up by the detectors of the LHC. But the director general of CERN, Rolf Heuer, urged caution, pointing out that the machine was still ramping up to full power. He predicted that by the end of 2012 the LHC would have collected sufficient data to definitively answer the Shakespearean question about the Higgs: 'to be or not to be'.[14]

Knowledge Is Power!

Far away in the heavenly abode of the great god Indra, there is a wonderful net which has been hung by some cunning artificer in such a manner that it stretches out infinitely in all directions. In accordance with the extravagant tastes of deities, the artificer has hung a single glittering jewel in each 'eye' of the net, and since the net itself is infinite in dimension, the jewels are infinite in number.

Francis H. Cook, *Hua-Yen Buddhism* (1977)[15]

If all the data generated in the LHC were to be stored, it would be the equivalent of saving millions of CD-ROMs full of information every second. This is far beyond the capacity of any current technology, so data collection presents the engineers at CERN with an enormous challenge. The electronics in the detectors are designed to be very selective about the data passed on for eventual storage. Of the billion or so collisions each second within ATLAS, the data from only a couple of hundred collision events will be deemed important enough to be worth storing. These are the collisions that have the telltale signs of interesting and possibly new physics. The system that performs the selection process and determines which information is discarded and which is stored for later analysis is called the trigger. The design of the trigger is absolutely critical. Without a well-thought-out trigger, most of the valuable data that everyone has worked so hard to generate would simply be thrown away. The trigger decision is based on the types of particle that have been detected. A good trigger is one that is fast, keeps all the interesting events and can be reprogrammed if the need arises. Good indicators of an interesting collision event might be the presence of electrons, muons, photons, particular types of jets or a signal that there is missing energy which indicates that neutrinos have been emitted. If the event passes these tests, it is sent on for data compression and further processing.

The primary selection is applied to the raw data coming out of the detector subsystems. The trigger electronics must decide very quickly which events are sufficiently interesting to pass along to the next stage of filtering. This reduces the total number of events from almost a billion to around two hundred per second. What remains is still a huge, but manageable data-handling task. Each proton–proton collision within ATLAS generates around 1.5 MB of data, to be stored for later off-line analysis. (This includes the data from the other twenty or so overlapping collision events produced in the same bunch crossing.) Altogether, around 300 MB of data from ATLAS will be stored every second – almost enough to fill half a CD-ROM. Each ten-hour run generates enough information to fill about 15,000 CD-ROMs. In a year this would produce a stack of CD-ROMs over seven kilometres high, even without their cases. If the data from the other detectors at the LHC are included, this figure doubles. Altogether the LHC will generate over 10 million gigabytes of information each year. Handling this enormous quantity of data has required the development of new collaborative methods of data processing and storage.

CERN has come a long way from the data processing of the past. In the 1960s, its main data network was pedal-powered. The data from the accelerators were recorded on magnetic tapes which were loaded into a basket on a bicycle and then rushed over to the computer centre by a physicist pedalling at top speed. Things are rather different today. Researchers around Europe and throughout the world need instant access to the experimental data from the LHC, so the data are not stored at a single site. A new distributed approach to information processing has been developed to cope with the unprecedented amount of data that will be generated. The data will be distributed around the world to over 140 computing centres in thirty-three countries.[16] This system is known formally as the Worldwide LHC Computing Grid, or more simply as the Grid. It is a service for sharing computer power and data storage capability throughout the globe over ultra-high-speed internet links. The aim is to turn the global network of computers into one vast computational resource. The primary back-up will be recorded on tape at CERN, and this will represent the Tier-0 centre of the Grid. After initial processing, the data will be distributed to a number of Tier-1 centres. These are large computing centres able to provide 24-hour a day support for the Grid and with sufficient storage capacity to hold a large proportion of the data. Tier-1 centres will then make the data available to smaller Tier-2 centres, and physicists will access the data through Tier-3 centres, which consist of computer clusters in university departments or individual PCs.

It is almost inevitable that fundamental research in physics leads to new technologies. In the words of the old adage, knowledge is power. During the course of our journey, we have seen many examples of technologies that have arisen from the investigations of physicists. These pioneers were driven primarily by pure curiosity, but modern life would be inconceivable without the tools, gadgets and devices that emerged from their research. Before Volta there were no batteries, and without Faraday's experiments there would be no electrical industry, so we would have none of the electrical appliances that we take for granted today. Without the electrochemistry of Humphry Davy there would be no sodium or potassium, but neither would there be aluminium cans or jet aeroplanes.

Maxwell's equations and Hertz's experiments led directly to the discovery of radio waves and the invention of the radio. J.J. Thomson discovered the electron using cathode-ray tubes that would become an integral component of the television set. The insight provided by quantum mechanics was necessary for the development of a whole range of modern technologies, such as lasers and the silicon chips that are found in digital cameras, computers, DVD players, mobile phones and satellite navigation systems. Even particle accelerators and superconducting magnets are no longer confined to physics laboratories. They are found in far greater quantities in hospitals, helping to save many lives. It is no exaggeration to say that the roots of every aspect of our technological society can be traced back to the musings of theorists and the probing of nature's secrets in research laboratories. The modest amounts that have been spent on fundamental research must rank as the most fruitful investments in history, and usually the dedicated individuals who contribute to these great breakthroughs leave the commercial exploitation to others, as their attention moves on to the next scientific puzzle. Their discoveries represent the greatest of all gifts to humanity.

This will surely continue. But the very nature of peering into the unknown means that it is all but impossible to predict what wondrous technologies will emerge from current cutting-edge research. Who knows whether grid computing or some other form of distributed computing will lead to benefits on the scale of the World Wide Web?

The Big Day Arrives at Last

On 13 December 2011, the media were eagerly awaiting an update from the physicists at CERN. The announcement was that the data from ATLAS and CMS ruled out the possibility of the Higgs in almost its entire possible mass range, but both detectors had glimpsed what might be the first signs of the Higgs with a mass close to 125 GeV. Much more data was still required to determine whether this was just a statistical blip or whether it was the real thing.

The big day finally arrived on 4 July 2012. Appropriately, the announcement was made in a webcast streamed live to the world. The spokesperson for the CMS team, Joe Incandela, gave the first presentation. He was followed by Fabiola Gianotti, on behalf of the ATLAS collaboration. The results from both detectors independently showed that a new particle was being detected with a mass in the range 125–126 GeV. With excitement mounting, at the end of the two hour-long talks, director general Rolf Heuer summed up with the words: 'If I was a layman, I would say I think we have it – you agree?', and the lecture theatre erupted with cheers and rapturous applause. Peter Higgs, who was in the audience, took off his glasses and appeared to brush away a tear.

Heuer went on to say: 'We have reached a milestone in our understanding of nature. The discovery of a particle consistent with the Higgs opens the way to more detailed studies, requiring larger statistics, which will pin down the new particle's properties, and is likely to shed light on other mysteries of our universe.' [17]

Finding the Higgs represents the most sensational discovery in particle physics in the four decades since the discovery of charm and the November Revolution of 1974. It puts the final piece in the jigsaw of the Standard Model and closes an important chapter in the history of physics. It is a clear sign that theorists are following the true path to a complete understanding of the forces of nature. But this is not the end of the story. Physicists are perfectionists. They want to resolve all the unanswered questions. The arrival of the Higgs opens a new door onto the ultimate structure of the universe, offering the possibility of exploring a completely new regime of physics beyond the Standard Model. The physical properties of the Higgs, such as how it decays and how strongly it interacts with all the other particles, will provide important new tests of the accuracy of the theory, and any deviations from the predictions of the Standard Model will be vital clues to an even better theory of the forces of nature.[18]

Beyond the Standard Model

Since the upheaval of the November Revolution of 1974 and the consolidation of the Standard Model in its aftermath, particle physics theory has remained ahead of accelerator experiments. Researchers have known what to expect when conducting their experiments, and there have been no really startling surprises. The detailed predictions of the Standard Model have been confirmed again and again, and not a single experiment has produced results that contradict it. The grand synthesis has truly stood the test of time. The LHC will enable physicists to reach an energy regime well beyond anything previously possible. For the first time in the era of particle colliders, the LHC will be able to probe beyond the now well-established physics of the Standard Model. It will almost certainly push the theory beyond its breaking point and reveal new physics. It will lead physicists on a journey into terra incognita, a realm in which they will soon come face to face with the unknown.

Now that the Higgs has been found, the aim of the LHC will be to find results that disagree with the Standard Model. The Standard Model must have a breaking point. As sensational as the theory is, it leaves far too many loose ends to be the ultimate answer, so at some energy scale, which physicists assume the LHC will reach, a new theory will be necessary. When an even better theory is discovered, the Standard Model will fit neatly within it. It must do, because otherwise it wouldn't work so well. (This is similar to how QED fits within the GWS electroweak theory.)

According to the Standard Model, particles possess mass because of the effect of the Higgs force. However, it does not explain why the different types of particles have the masses that they have: why, for instance, a muon is 207 times heavier than an electron, or why the mass of the top quark is so enormous compared with the masses of each of the other quarks. Neither can it explain why there are three generations of elementary matter particles.

Another question that the Standard Model has been unable to resolve is the embarrassing fact that most of the matter in the universe appears to consist of an unknown substance. This substance does not emit light and is only known through its gravitational influence on other matter. For this reason it is known as dark matter. It is usually assumed that dark matter is composed of huge quantities of a very weakly interacting relic particle that was produced in prodigious amounts soon after the origin of the universe in the Big Bang. If this is correct, then this particle may turn up in the Large Hadron Collider.

Physicists yearn for theories even more beautiful and all-encompassing than the Standard Model. They long for theories built around even higher degrees of symmetry. One feature of the Standard Model that is perhaps not as satisfying as it might be is that the matter particles have a quite distinct status from the force-carrying particles. Fundamental particles fall into two separate classes that display completely different behaviour. One class of particles form the constituents of matter, while the other class mediate the forces that hold matter together. Matter-forming particles, such as the electron, obey the exclusion principle, which means that each one must be in a different wave state. This curious property is shared by all the fundamental particles from which matter is formed, including protons, neutrons and quarks. In fact, it is this property that enables them to condense into matter. Collectively, these particles are known as fermions. By contrast, particles whose exchange produces a force, such as photons, behave in a completely different way. They like nothing more than to all exist in the same wave. This is the basis of a laser: a laser beam is composed of vast numbers of photons that are all simultaneously forming a single wave. Particles that behave in this way are collectively known as bosons.

This is why the Standard Model table of elementary particles comes in two parts. The first consists of the fermions – the three generations of four matter particles; the second consists of the bosons or exchange particles – the octet of gluons plus the quartet of particles that mediate the electroweak force and the Higgs. Quantum mechanics explains why the two classes of particle behave so differently. But it would be nice if there were some sort of symmetry that related the two types of particle. The matter and force-carrying particles could then be paired up, with a force-carrying particle for each matter particle and vice versa. This is exactly the sort of deep relationship that physicists strive for, a unity at the heart of matter that leads to a more profound understanding of the universe. Its discovery would represent a major step forward towards the goal of a total unification of all the forces and all the particles within a single theory. Physicists have given a name to this type of symmetry: supersymmetry.

Winos and Squarks

> After the division, the two parts of man, each desiring the other half, came together, and throwing their arms about one another, entwined in mutual embraces, longing to grow into one … So ancient is the desire of one another which is implanted in us, reuniting our original nature, seeking to make one of two, and to heal the state of man.
>
> Plato, *Symposium*[19]

Supersymmetry is a novel type of symmetry, quite different from the symmetries previously encountered by theoretical physicists. It is a symmetry between matter-forming particles, such as electrons and quarks, and force-mediating particles, such as photons, gluons and the Higgs. Supersymmetry unites the matter particles and the force particles in a warm, mutual embrace. It is as though the sturdy masculine matter particles have been entwined with the feminine force-carrying particles that radiate their charms and bind the particles together. If the universe really is governed by supersymmetric laws, then each type of matter particle has a supersymmetry partner, or superpartner, that is a force-mediating particle; conversely, each force-mediating particle has a superpartner that is a matter particle.

However, the matter particle and the force-carrier in a supersymmetry pairing must have exactly the same charges, and this is not the case for the particles that are currently known to physicists, so the known particles cannot be paired up in this way. So if supersymmetry is a symmetry of the real world, it would imply the existence of many different new particles. There would be a new superpartner particle for each known particle. Physicists have names for all these particles, even though none has yet been seen. The names of the matter particles that partner the known force-carriers are produced by adding the suffix 'ino' to the end of the name of each force-carrier. For instance, if the universe really is supersymmetric, the photon has a superpartner that is known as a photino. The superpartners of the W and Z particles are Winos and Zinos. Similarly, the superpartners of the gluons are gluinos, and the superpartner of the Higgs is the Higgsino. The names of the force-carrying particles that partner the known matter particles are derived by adding the letter s, which stands for 'super', to the start of the name of the matter particle. For instance, the superpartner of the electron is known as the selectron, and the superpartners of the quarks are known as squarks.

Unbroken supersymmetry would imply that particles have exactly the same mass as their superpartners – which is clearly not the case. The masses of all the new superpartner particles must be much greater than the masses of the known particles, otherwise these particles would already have been seen in previous generations of particle accelerators. The only conclusion is that, if supersymmetry does play a role in the real world, then it must be a spontaneously broken symmetry, just like the electroweak symmetry. In fact, many theorists believe that the breaking of supersymmetry may be related to the breaking of the electroweak force. If so, it is reasonable to expect the energy at which supersymmetry is broken to be similar to the energy at which the electroweak force is broken, and this would imply that the lightest of the superpartner particles would then have a similar mass to that of the Higgs. If this is true, then there could be a whole host of new particles awaiting discovery by the LHC. However, the arguments in favour of supersymmetry are not as compelling as they were for the Higgs.

But who knows – evidence for supersymmetry may yet appear at the LHC. Its discovery would be an amazing experimental revelation, confirming one of the most daring predictions in the history of physics. Furthermore, the lightest of the superpartners would be a completely stable particle, produced in prodigious quantities in the early universe. Even now, space would be filled with vast quantities of the particle. If this hypothesis is correct, then the origin of the dark matter – which constitutes most of the matter in the universe – might be uncovered at the LHC. A decade from now, in another edition of this book, there may be another chapter describing these startling discoveries and how they have helped to build an even more beautiful theory of the cosmos and its constituents.

The Ultimate Prize

> Love is simply the name for the desire and the pursuit of the whole.
>
> Plato, *Symposium*[20]

Supersymmetry is just one of the ideas that have sprung from the fertile minds of theorists since the Standard Model was formulated in the 1970s. The ultimate goal of theorists is to find a single theory that encompasses the whole of physics and answers all our fundamental questions about the universe and the material within it. This is tremendously

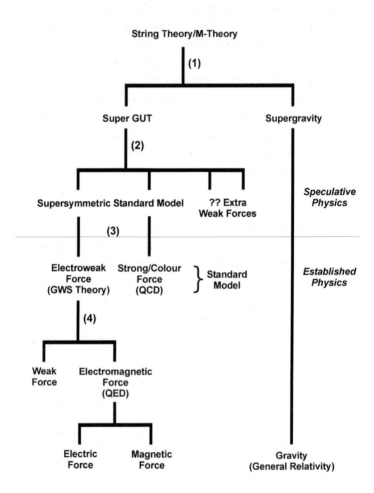

ambitious, but in recent decades there have been major advances that appear to be heading in the right direction. Figure 79 illustrates how this total unification might come about. The bottom half of the diagram represents the well-established physics that has been described in this book, which is attested by mountains of experimental evidence. Our whole understanding of physics is now contained within our best theory of gravity, which is Einstein's general relativity, plus the Standard Model, which is made up of the GWS electroweak theory and the theory of the strong or colour force (QCD). But there are still loose ends that need to be woven together to produce the ultimate unified theory, and the top half of the diagram represents the best attempts by physicists to complete the tapestry.

By the early 1970s it had become clear that the strong force and the electroweak force can both be modelled by Yang–Mills theories, but with different symmetry groups. The next step towards total unification was to explore the possibility that at extremely high energies both the strong force and the electroweak force merge into one force, and that this force could be represented by a Yang–Mills theory with a single,

---

Figure 79 *(opposite)* The ongoing attempts to unify all the forces of nature. The bottom half of the diagram depicts the established physics described in this book. The top half of the diagram depicts further theoretical steps towards total unification. The numbers indicate four major stages of symmetry breaking. (4) The Higgs mechanism, which breaks the symmetry between the electromagnetic and weak forces. (3) Supersymmetry breaking. (2) The breaking of the symmetry of a grand unified theory to the symmetry of the standard model. The extra weak forces indicated here are possible additional forces that are currently unknown, but may be discovered at the LHC. (1) The breaking of the symmetry of string theory to the symmetry of a superGUT (supersymmetric grand unified theory) and supergravity (supersymmetric quantum theory of gravity).

larger unified symmetry group. This would explain why the components of the Standard Model fit together in such a harmonious way. Such theories are known as Grand Unified Theories, or GUTs. (The supersymmetric versions are, as you might expect, known as superGUTs.) To get from the GUT to the Standard Model requires the symmetry of the GUT to be spontaneously broken by a similar mechanism to the Higgs mechanism that breaks the symmetry of the electroweak force to produce separate electromagnetic and weak forces. This symmetry breaking must occur at ultra-high temperatures, much higher than the trillion degrees at which the electroweak symmetry is broken. One implication of Grand Unified Theories is that matter is ultimately unstable. They suggest that very occasionally we should expect a proton within one of the atoms in our bodies to disintegrate. To test this, physicists have been peering into huge tanks of ultrapure water looking for the extremely rare decays of a proton. These experiments have now been in progress for several decades, but the protons seem to be hanging on, without any sign of wanting to decay.

Since around 1984, studying the theoretical unification of all the forces, including gravity, has become a mainstream activity for many physicists and mathematicians. A plausible framework for tying the whole of physics together has emerged. This theory is known as string theory. Traditionally, particles have been thought of as point-like entities. The novel feature that string theory brings to a unified description of matter is the idea that matter might be built from one-dimensional entities – or strings – that can vibrate in many different ways, the different modes of vibration representing different particles. For instance, in one mode of vibration the string might behave like a quark, in another mode like an electron and in a third mode like a photon. In recent decades string theory has been explored as a potential ultimate theory of physics. It has many of the components that are essential in a unified description of the universe. It contains a quantum theory of gravity and describes the other forces of particle physics within a grand unified theory, and it is also, necessarily, supersymmetric. These seductive features have enticed many of the world's leading physicists to dedicate their lives to unravelling the deep and still mysterious mathematical and physical implications of string theory.

Theorists would like their ultimate theory of the universe to be in some way unique. But by the middle of the 1990s it appeared that there were five different string theories to choose from, which greatly weakened the claim that any of them could be the ultimate theory. In 1995, the second superstring revolution was triggered by Ed Witten, the world's leading string theorist, who realised that the five string theories represent different parts of an all-embracing new theory that he named M-theory. This theory has still not been fully formulated. The 'M' in the name encapsulates the current ignorance about the true nature of the theory. It simultaneously stands for 'membrane', 'matrix', 'magical', 'mysterious' and 'miracle'.

Currently, there is no experimental evidence to support the theories represented in the top half of Figure 79. Their plausibility rests on the beautiful abstract tapestries that have been woven with them. They tie up the loose ends of the Standard Model and general relativity in a mathematically consistent way, which is no mean feat, but only time will tell whether they do this in a way that accurately represents reality.

The LHC promises to shed light on the earliest moments of the universe and the Big Bang fireball in which it came into existence. It might enlighten us about the large-scale composition of the universe, but it will certainly enable us to delve deeper into the fundamental structure of matter and understand the forces that hold matter together. This knowledge will give us more insight into the universe and how it came to have the subtle balance of properties that has allowed life to emerge and evolve into sentient beings who can reflect on these ideas. The LHC will help us to contemplate the universe and understand our true place within it.

The Higgs particle is the first great particle-physics discovery of the third millennium. It is definitive proof that we really do live within a 'cosmic superconductor'. Peter Higgs, the symmetry breaker, has reached his eighties. Now that the Higgs particle has finally arrived, he can surely expect a speedy telegram from the King of Norway informing him that he has been awarded the Nobel Prize in Physics.

Appendix

LANDAU AND HIGGS

This appendix uses some simple algebra to explain Landau's theory of phase transitions and how it was adapted to particle physics by Higgs. Landau recognised that a phase transition is associated with a change in symmetry, such that symmetries which hold above the transition temperature may be lost below the transition temperature. This loss of symmetry is equivalent to a gain in order, and Landau suggested that there must be a parameter that measures the order. Above the transition temperature the order parameter is zero, as the system is completely disordered, but below the transition temperature the order parameter takes a non-zero value which is a measure of the order that has spontaneously arisen.

These ideas are illustrated in Chapter 7 in terms of a toy model consisting of a collection of atoms arranged regularly in a plane. Each atom has a small magnetic field that is constrained to point in either of two directions: upwards or downwards. Both directions are completely equivalent, but the energy of two neighbouring atoms is lowered slightly if their magnetic fields point in the same direction, so there is a tendency for them to align. Above the transition temperature, random thermal oscillations of the atoms will disrupt the alignments. Below this temperature the thermal oscillations will be too weak to disrupt the alignments and the magnetic fields will align.

In the case of the magnet, Landau's order parameter is called the magnetisation. It is equal to the proportion of atoms whose magnetic fields are pointing upwards minus the proportion that are pointing downwards. Above the transition temperature the magnetisation is zero; below the transition temperature it is non-zero. The next step in Landau's analysis was to write down an expression for the energy of the system near the transition temperature in terms of the magnetisation, M. Close to the transition temperature M must be small, as it is zero above the transition temperature, so Landau assumed that the energy could be represented as a polynomial expansion in M. Now, the toy model is completely symmetrical with respect to the up and down directions of the magnetic fields, so the energy cannot depend directly on M; if it did, an upward magnetic field would contribute $+M$ to the energy and a downward magnetic field would contribute $-M$ to the energy, and that would contradict the assumption that the system is completely symmetrical with respect to the two directions. In other words, the expression for the energy must respect the symmetry of the system.

The energy cannot depend on any odd power of M. For instance, if the energy were proportional to M^3, then the contribution of an upward magnetic field would be equal to minus the contribution of a downwards magnetic field. The same is true for all odd powers of M, because -1 raised to an odd power is equal to -1. Conversely, if the energy were proportional to an even power M, then both the upward and downward magnetic fields would make the same contribution to the energy, because -1 raised to an even power is equal to $+1$. The expression must therefore only contain even powers of M. Furthermore, all powers of M higher than the fourth power can be disregarded, as they are insignificant when M is small. (Near the transition temperature, the magnetisation is much less than 1, so successive powers of M get rapidly smaller as the power increases. For instance, if $M = 0.1$, then $M^2 = 0.01$, $M^4 = 0.0001$, $M^6 = 0.000\,001$, and so on.) This leaves just two terms in the expansion, terms proportional to M^2 and M^4. The result is the following expression for the energy E in terms of the magnetisation at temperatures close to the phase transition:

$$E = \alpha M^2 + \beta M^4$$

This is the most general expression that satisfies the criteria set out in Landau's argument.

Physical systems always tend to fall into the lowest energy state available to them, and a reasonably large system will reach this lowest energy state very rapidly. So this expression is extremely important as it determines the state that the system will find itself in. Landau's interpretation of the phase transition is based on the shape of the graph of the energy when plotted against the magnetisation. Now, the expression for the energy contains two unknown terms, which we are calling α and β. Landau reasoned that β must be positive, because otherwise there would be no minimum energy, and this clearly could not represent a physical system. This implied that there were just two possibilities for the shape of the energy graph.

If α were positive, the graph would look like Figure A1, where the energy has a single minimum corresponding to the point where the magnetisation M is equal to zero. But if α were negative, the graph would look like Figure A2, where there are now two values of the magnetisation at which the energy takes its lowest possible value. These values of the magnetisation are labelled 't' and '$-t$'. If you are familiar with differential calculus, it is easy to check the form of these two graphs. The turning points of the graphs are the points at which the derivative of E with respect to M is equal to zero:

$$\mathrm{d}E/\mathrm{d}M = 2\alpha M + 4\beta M^3 = 0$$

If both α and β are positive, there is just a single minimum, at $M = 0$. However, if α is negative and β is positive, there

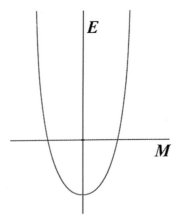

Figure A1 Graph of the energy E plotted against the magnetisation M, where the energy is given by the expression $E = \alpha M^2 + \beta M^4$ and both α and β are positive. There is one minimum, at $M = 0$.

are three turning points, at $M = 0$, $M = -\sqrt{(|\alpha|/2\beta)}$ and $M = \sqrt{(|\alpha|/2\beta)}$. The turning point at $M = 0$ is a maximum and the other two are minima.

Landau reasoned that the first graph (Figure A1) must represent the energy just above the transition temperature, and the second graph (Figure A2) must represent the energy just below the transition temperature. In other words, α is not a constant: it must vary with temperature. Above the transition temperature α is positive, below the transition temperature α is negative, and at the transition temperature α is exactly equal to zero.

Landau's description of the phase transition was that above the transition temperature the system is in its lowest energy state, which is the minimum of the first energy graph. At this point the magnetisation is zero, which represents a state in which the magnetic fields are oriented at random with respect to the up–down axis, so that there are equal numbers pointing

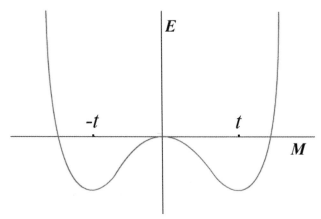

Figure A2 Graph of the energy E plotted against the magnetisation M, where the energy is given by the expression $E = \alpha M^2 + \beta M^4$, and β is positive but α is negative. There are two minima, at $M = -t$ and $M = t$.

upwards and downwards. However, as the temperature falls, the shape of the graph changes, and below the transition temperature there are two minima, and the system must fall into one of these minima. They are both equivalent, so the choice is made completely arbitrarily by random fluctuations within the system as the temperature falls below the transition temperature. At one minimum all the atomic magnetic fields will be pointing upwards; at the other they will all be pointing downwards. In either of these states the system is now ordered, with all the magnetic fields aligned and the up–down symmetry is lost – or at least hidden.

Landau's expression relating the energy to the order parameter is applicable to a whole range of phase transitions, and the energy graphs generalise in a natural way to higher-dimensional systems. For instance, if the magnetic fields were free to point in any direction in a plane rather than being constrained to lie along a single axis, then below the transition temperature the energy graph would look like Figure 65 (page 317). To form this graph, just rotate the second graph shown above around the energy axis. This is known as a Mexican hat diagram. The lowest energy states now form a circle within the brim of the hat. At each point on this circle the energy of the magnetic system is as low as possible.

Peter Higgs adapted Landau's model to particle physics. Plate 9 is a photograph of Higgs standing in front of a blackboard. The Mexican hat diagram has been drawn on the left of the blackboard. After reading the following, you might recognise some of the algebra on the blackboard as well. In quantum field theories, each type of fundamental particle is associated with a quantum field that permeates the whole of space. All these fields fluctuate at each point in space, and the excitations of the various fields are what we interpret as particles. In the model constructed by Higgs there is a field, 'phi' (ϕ) that plays the role of the order parameter in Landau's theory. The energy of the ϕ field is usually represented by the letter V, where

$$V = \alpha\phi^2 + \beta\phi^4$$

This is exactly the same form as the energy in Landau's theory of phase transitions. The only difference is that here ϕ has been swapped for M. (This expression is written on the blackboard behind Higgs, but ϕ^2 is written as $\phi^*\phi$ and ϕ^4 is written as $(\phi^*\phi)^2$, because ϕ is a complex number.) In quantum field theory, excitations of the field are interpreted as particles and β is interpreted as the strength with which the ϕ particles interact with one another. These interactions may be represented as a Feynman diagram, as in Figure A3. At the bottom of the diagram two ϕ particles approach each other, at the centre they interact with strength β, then at the top they recede from each other. (The two incoming ϕs and the two outgoing ϕs are the four ϕs in the expression ϕ^4.) The other term in the expression for the energy, $\alpha\phi^2$, is the energy cost of producing a ϕ particle. In quantum field theory, α is interpreted as the mass of the ϕ particles.

In addition to the field ϕ, Higgs's model includes a field, A, corresponding to a force-mediating particle that would behave like a photon. The interaction between the A particles and the ϕ particles produces an additional contribution to the energy of the form $e^2A^2\phi^2$, where e is the electric charge of the ϕ particle. This interaction is represented in Figure A4 as a Feynman diagram. It represents a process in which an A particle and a ϕ particle approach each other, interact with strength e^2 and then recede from each other.

439

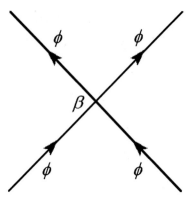

Figure A3 As a Feynman diagram, the quartic term $\beta\phi^4$ in the energy represents two incoming ϕ particles that interact with strength β and then recede from each other.

Higgs proposed that immediately after the Big Bang the universe went through a dramatic phase transition. At extremely high temperatures in the very early universe, the parameter α would be positive. The lowest energy configuration would then be one in which the ϕ field was zero. But, in accordance with Landau's theory, as the temperature dropped the value of α decreased until, below a transition temperature of perhaps a trillion degrees, it became negative. The energy graph would then look like the Mexican hat diagram; below the transition temperature the lowest energy configurations of the ϕ field form a circle inside the brim of the hat. The ϕ field now loses energy and the system falls randomly to one of the points on this minimum-energy circle. In other words, ϕ now takes a non-zero value that minimises the energy. The result is that the ϕ field is no longer zero, even in completely empty space, just as the magnetic field in a permanent magnet is not zero. As a result of the phase transition, the symmetry of the system has been spontaneously broken.

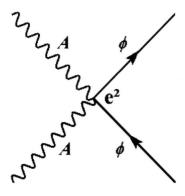

Figure A4 As a Feynman diagram, the expression $e^2A^2\phi^2$ represents an incoming A particle and an incoming ϕ particle that interact with strength e^2 and then recede from each other

The quantum field will oscillate around the minimum-energy field. To make this explicit, we need to define a new field H that is zero at the energy minimum and examine the oscillations and excitations of this field. If we represent this minimum energy value of the field ϕ as t (which is equal to $\surd(|\alpha|/2\beta)$), ϕ can be rewritten as

$$\phi = t + H$$

where t is now interpreted as the value of the constant background ϕ field that permeates empty space, and H is the Higgs field. Expanding Landau's expression for the energy in the ϕ field in terms of this new field will explicitly show the effect of the symmetry breaking. The result is a constant term, which particle physicists usually disregard as it simply shifts the energy scale; then there is a quadratic term and a quartic term, just as there was for the original ϕ field, but now there is also a term proportional to H^3, so the energy is not symmetrical under $H \rightarrow -H$. This symmetry has been lost in the new low-energy configuration.

But more important is the effect this symmetry breaking has on the photon-like A particle. The energy of the interaction between the ϕ particle and the A particle is now represented by the term $e^2 A^2 \phi^2$, as shown in Figure A4. Rewriting this in terms of the Higgs field gives

$$e^2 A^2 (t + H)^2$$

which expands to

$$e^2 t^2 A^2 + 2e^2 t H A^2 + e^2 H^2 A^2$$

Here the second and third terms represent interactions between the Higgs particle and the A particle, but the first term is the one we are most interested in. It corresponds to the interaction between the A particle and the background ϕ field, which takes the value t. It is a quadratic term in A multiplied by a constant, $e^2 t^2$. This takes the form of a mass term for the A particle that has arisen from its interactions with the background ϕ field. In the symmetric phase at temperatures above the phase transition, the force-carrying particle A had no mass, but because of its interactions with the background ϕ field it now has a mass equal to $e^2 t^2$. This is the magic of the Higgs mechanism. The symmetry of the force has been spontaneously broken, and the photon-like particle that mediates the force has been transformed into a massive particle. The force will now be a feeble short-range force, like the weak force.

Notes

INTRODUCTION

1. The Higgs particle is often referred to as the Higgs boson. There are two fundamental categories of particles, known to physicists as fermions and bosons. Fermions are the particles, such as protons and electrons, from which matter is composed. Bosons are particles that are passed between other particles to produce forces. For example, the exchange of fundamental particles of light, known as photons, is what produces the electromagnetic force. The Higgs boson is responsible for the force that gave this book its title.

2. Protons are subcomponents of atoms and are found in the nucleus of all atoms. 'Hadron' is a collective name for nuclear particles such as protons. The reason that the LHC is not simply called the Large Proton Collider is that it is a multi-purpose machine and will also be used for experiments in which other particles, such as the nuclei of atoms of lead, are collided.

1 SEEING THE WORLD THROUGH KALEIDOSCOPE EYES

1. Colin A. Ronan, 'The fundamental ideas of Chinese science', chapter 10 of *The Shorter Science & Civilisation in China*, vol. 1 (Cambridge University Press, 1978). Ronan's five-volume work is a partial abridgement of Joseph Needham's original monumental text. From p. 164 onwards, this chapter describes the fundamental differences between the causal philosophies of the Greek philosophers and the organic worldview of Chinese science.

2. Morris Kline, *Mathematics in Western Culture*, revised edn (Penguin, 1987), p. 371.

3. The sum of the angles of a triangle is only 180° in flat or Euclidean space. The non-Euclidean geometries discovered in the nineteenth century are based on different sets of axioms, and for those geometries this result does not hold.

4. Bertrand Russell, *Mysticism and Logic And Other Essays* (Longman, 1919), p. 60.

5. Alfred North Whitehead, *Process and Reality*, 2nd revised edn (Macmillan, 1979), p. 39.

6. Plato, *Timaeus and Critias*, 17a.

7. It was the desire to produce a functional geometrical model of the cosmos in line with Plato's ideas that led Plato's pupil, the geometer Eudoxus of Cnidus, to develop a theory of the movements of the

planets in which they were assumed to be attached to transparent, concentric spheres.

8. Euclid's *Elements* are the elements of geometry and should not be confused with the elements of alchemy: earth, air, fire and water.

9. To modern mathematicians, a regular polyhedron is the surface of any of the solids that Plato had in mind. It is a two-dimensional surface formed of polygonal faces that encloses a three-dimensional region of space.

10. Perhaps the reason that Plato did not refer to these minuscule regular solids as atoms is because the word 'atom' means 'indivisible', and the purpose of Plato's theory was to explain the transmutation of matter in terms of the rearrangement of the component faces of the regular solids.

11. Kepler's fascinating life story forms the centrepiece of Arthur Koestler's wonderful account of the early history of celestial mechanics, *The Sleepwalkers: A History of Man's Changing Vision of the Universe* (Penguin, 1989).

12. The Ptolemaic system was based on models of the cosmos developed by a succession of Greek mathematicians, including Eudoxus, Apollonius and Hipparchus.

13. Johannes Kepler, *The Six-Cornered Snowflake*, translated by Jacques Bromberg (Paul Dry Books, 2010).

14. Richard P. Feynman, Robert B. Leighton and Matthew Sands, *The Feynman Lectures on Physics*, vol. 1 (Addison-Wesley, 1963), p. 14.

15. The centres of these six circles are the vertices of a regular hexagon centred on the central circle.

16. The cuboctahedron is known as a semi-regular polyhedron because, unlike the Platonic solids, it has faces of two types – triangles and squares.

17. Chemists call this packing face-centred cubic.

18. It has proved much easier to find the densest sphere packings in spaces of higher dimensions, such as 4, 8 and 24; see John Conway and Neil Sloane, *Sphere Packings, Lattices and Groups*, 2nd edn (Springer-Verlag, 1993), p. 22.

19. Simon Singh, 'A new solution to the mathematical problem of packing spheres brings with it more problems', *Daily Telegraph*, 13 August 1998. Reproduced at www.simonsingh.net/media/articles/maths-and-science/how-to-stack-oranges/.

20. George G. Szpiro, 'But is it really a proof?', chapter 13 of *Kepler's Conjecture: How Some of the Greatest Minds in History Helped to Solve One of the Oldest Maths Problems in the World* (Wiley, 2003), p. 201.

21. Quoted in Richard Morris, *The Last Sorcerers: The Path from Alchemy to the Periodic Table* (Joseph Henry Press, 2003), p. 58.

22. Eric R. Scerri, *The Periodic Table: Its Story and Its Significance* (Oxford University Press, 2007), p. 30.

23. Jean-Baptiste Delambre, 'Notice sur la vie et les ouvrages de M. le Comte J.-L. Lagrange', in *Œuvres de Lagrange*, vol. 1, edited by J.A. Serret (Gauthier-Villars, 1867), p. xl. Quoted by Paul Strathern in *Mendeleyev's Dream: The Quest for the Elements* (Penguin, 2000), p. 241.

24. Flann O'Brien, *The Third Policeman* (Paladin, 1988), p. 86.

25. Morris, *The Last Sorcerers*, p. 130.

26. Strathern, *Mendeleyev's Dream*, p. 262.

27. Patience is the British name for solo card games; the US equivalent is solitaire.

28. Strathern, *Mendeleyev's Dream*, p. 286.

29. John Emsley, *Nature's Building Blocks: An A–Z Guide to the Elements* (Oxford University Press, 2001), p. 521.

30. Mendeleyev also correctly predicted the existence of the elements hafnium and technetium, which were not discovered until after his death. Technetium, Mendeleyev's eka-manganese, has a half-life of 4.2 million years. It is extremely rare as a naturally occurring element on Earth, being found as a fission product of uranium. It was discovered in 1937 when it was artificially produced in the laboratory. Several chemists claimed the discovery of hafnium, but were later shown to be in error. It was eventually isolated in 1923 by Dirk Coster and Gyorgy von Hevesy working at the Niels Bohr Institute in Copenhagen; they named the element after Hafnia, the Latin name for Copenhagen.

31. Jorge Luis Borges, 'The Plot', in *The Aleph and Other Stories* (Penguin, 2004), p. 157.

32. Richard Gregory, *Mirrors in Mind* (Penguin, 1997), p. 182.

33. The best account of the lives of Abel and Galois is in Simon Singh, *Fermat's Last Theorem: The Story of a Riddle That Confounded the World's Greatest Minds for 358 Years* (Fourth Estate, 2002).

34. This statement is attributed to Weyl by Richard Feynman in Feynman *et al.*, *The Feynman Lectures in Physics*, vol. 2, section 52–1.

35. The 'group' in group theory should not be confused with the word as used by chemists, who use it to refer to the elements in the same column of the periodic table.

36. The tetrahedra used to generate these kaleidoscopes are not regular tetrahedra.

37. To mathematicians, a 'honeycomb' is more general than a hexagonal cell structure made by bees: it is any space-filling collection of

polyhedra. More formally, a honeycomb is an n-dimensional space-filling collection of n-dimensional polytopes.

38. To form Kepler's packing, place a sphere at each vertex of the honeycomb and then place spheres at the centre of each of the cuboctahedra in the honeycomb.

39. Many of the symmetries that play a role in the fundamental laws of nature are continuous symmetries. A circle is an example of an object that is symmetrical under a continuous group of symmetries. A hexagon is symmetrical only under rotations through six different angles, the multiples of the 60° rotation, but a circle is symmetrical under a continuous range of rotations through any angle around its centre between 0° and 360°.

40. This puzzle was devised by David Wells. It is Puzzle 491, 'Mystic Square', in his *The Book of Curious and Interesting Puzzles* (Dover, 2006), and was originally published in *SymmetryPlus* magazine (a magazine for school pupils published by the Mathematical Association), no. 6, Summer 1987, p. 7.

41. We have all experienced the sensation of sitting in a train in a station when a train on the next track begins to pulls away – for a moment we are unsure whether it is our train that is moving or the other train.

42. Noether's result also supplies the answer to the question of what symmetry the conservation of electric charge corresponds to. The answer is more abstract than the previous examples: it is the symmetry of a circle, but not one that exists as part of physical space. It is the existence of this symmetry in nature that gives rise to the electromagnetic force.

43. Einstein considered symmetry to be such an important part of his theories that he later regretted using the name 'relativity' for his most famous theory, saying that 'invariance theory' would have been a much more appropriate name.

44. Arthur I. Miller, *Einstein, Picasso: Space, Time, and the Beauty that Causes Havoc* (Basic Books, 2002), p. 71.

2 UNITY IS STRENGTH

1. Keynes Collection, Kings College Cambridge, MS 130.4, p. 10. Quoted in Richard S. Westfall, *Never at Rest: A Biography of Isaac Newton* (Cambridge University Press, 2000), p. 154.

2. Details of the life and career of Gilbert may be found at www.magnet .fsu.edu/education/tutorials/pioneers/gilbert.html.

3. It is purely a matter of convention that the charge on the electron is defined to be negative and the charge on the proton is defined to

be positive. Of course, once the sign of the charge of one particle has been chosen, this determines the sign of the charge on the other. Our convention means that the direction of positive current flow is actually opposite to the direction of the flow of the negatively charged electrons that produce the current.

4. It was the French engineer Petrus Peregrinus who, noting this phenomenon in the thirteenth century, was the first to use the terms 'north pole' and 'south pole' for the ends of the magnet that point towards the Earth's north and south poles, respectively.

5. Franklin also systematised the laws of electrostatics, defined the conventions for naming positive and negative electric charges, and enunciated the law of conservation of charge.

6. Edmund Clerihew Bentley invented the verse form now known as the clerihew. The very first clerihew composed by Bentley as a schoolboy was this four-line verse poking fun at Humphry Davy.

7. Humphry Davy, *Researches Chemical and Philosophical Chiefly Concerning Nitrous Oxide, or Dephlogisticated Nitrous Air, and Its Respiration* (J. Johnson, 1800), p. 556.

8. It is now known that inhalation of nitrous oxide can have serious side-effects. It interferes with the metabolism of vitamin B_{12}, cobalamin. The long-term inhalation of nitrous oxide and other gases may have contributed to Davy's early death in Geneva in 1829 at the age of fifty.

9. John Buckingham, *Chasing the Molecule: Discovering the Building Blocks of Life* (Sutton Publishing, 2005), p. 27.

10. Buckingham, *Chasing the Molecule*, p. 84.

11. Davy would go on to discover three more elements: magnesium, boron and barium.

12. For more details of Faraday's life, work and legacy, see Iwan Rhys Morus, *Michael Faraday and the Electrical Century* (Icon, 2004), p. 11.

13. Michael Faraday, 'First letter to Benjamin Abbott, member of the City Philosophical Society, 12 July 1812', in *The Life and Letters of Faraday*, vol. 1, 2nd revised edn (Cambridge University Press, 2010), p. 17.

14. Malcolm Longair, *Theoretical Concepts in Physics: An Alternative View of Theoretical Reasoning in Physics*, 2nd edn (Cambridge University Press, 2003), p. 88.

15. It was published in three parts as 'Historical sketch of electromagnetism', *Annals of Philosophy*, vol. 18, pp. 195 and 274 (1821); vol. 19, p. 107 (1822).

16. Longair, *Theoretical Concepts in Physics*, p. 83.

17. In an alternative arrangement of his apparatus, Faraday fixed the magnet in position with one pole projecting above the surface of the

mercury in the beaker, but allowed the end of the wire attached to the cork floating on top of the mercury to move. This time, when the circuit was closed the cork and the attached wire rotated around the magnet.

18. We now know that the magnetic field of a bar magnet results from the alignment of the spins of iron atoms, and that it is the rotation of the electric fields in these atoms that produces the magnetic fields that they exhibit.

19. Longair, *Theoretical Concepts in Physics*, p. 85.

20. Ian Mortimer, *The Time Traveller's Guide to Medieval England* (Vintage, 2009), p. 5.

21. Coulomb used a torsion balance to investigate the electrostatic and magnetic forces, and published the results in his seven-volume *Mémoire sur l'Électricité et le Magnétisme*.

22. Longair, *Theoretical Concepts in Physics*, p. 88.

23. A magnetic monopole would be a particle carrying a magnetic charge, rather than an electric charge. The existence of such particles is predicted by some of the more speculative theories of particle physics, but none has ever been seen.

24. Faraday had performed experiments in which an electric field changed the polarisation of light. He therefore suspected that there was a link between light and electromagnetism. Maxwell visited Faraday on his deathbed and described his discovery to him.

3 THE DREAMS THAT STUFF IS MADE OF

1. Quoted in Abraham Pais, *Inward Bound: Of Matter and Forces in the Physical World* (Oxford University Press, 1988), p. 165.

2. Visible light is electromagnetic radiation whose wavelength is in the range that can be detected by the photosensitive pigments in our rods and cones, the cells that form the retinas of our eyes. These pigments are sensitive to electromagnetic radiation with a wavelength between about 400 and 700 nm (nm stands for nanometre; one nanometre is one-billionth of a metre, i.e. one-millionth of a millimetre). The colour we perceive is determined by the wavelength of the light. Light with a wavelength of 700 nm is perceived as red; light with a wavelength of 400 nm is perceived as violet. Between these wavelengths the light goes through a continuous spectrum of colour – the colours of the rainbow – from red, through orange, yellow, green, blue and indigo, to violet. This division of the spectrum of visible light into seven colours was first noted by Newton, though it is debatable whether the colour indigo can really be distinguished from blue and violet.

3. Quoted in Mary Elvira Weeks, 'Some spectroscopic discoveries', in *Discovery of the Elements*, revised edn (Kessinger Publishing, 2003), p. 262.

4. John Gribbin, *Stardust: The Cosmic Recycling of Stars, Planets and People* (Penguin, 2000), p. 53.

5. During the first twenty minutes or so after the Big Bang, the matter in the universe underwent nuclear fusion reactions. A quarter of all the hydrogen was transformed into helium, and trace amounts of elements 3 and 4, lithium and beryllium, were synthesised. By the end of this brief period of nuclear fusion the density of matter in the universe had fallen too low to support further nuclear reactions, no more of which occurred until the first stars formed.

6. The whole apparatus was enclosed in a glass vacuum tube as the alpha particles would have been absorbed on their passage through the air.

7. Quoted in David Wilson, *Rutherford: Simple Genius* (Hodder & Stoughton, 1983), p. 155. According to Wilson, it was published in the *McGill University News* in 1904.

8. Quoted in Steven Weinberg, *The Discovery of Subatomic Particles* (W.H. Freeman, 1990), p. 122.

9. H. Geiger and E. Marsden, 'The laws of deflexion of particles through large angles', *Philosophical Magazine*, vol. 25, p. 604 (1913).

10. Ernest Rutherford, 'The structure of the atom', *Philosophical Magazine*, vol. 27, p. 488 (1914).

11. Weinberg, *The Discovery of Subatomic Particles*, p. 127.

12. When electrons are accelerated they emit electromagnetic radiation. For instance, if an alternating current is passed through the aerial of a radio transmitter it will jostle the electrons up and down and cause them to emit radio waves. Exactly the same effect causes the instability of the 'classical' Rutherford atom.

13. Quoted in Claes Johnson, *Dr Faustus of Modern Physics* (Icarus eBooks, 2011), p. 102.

14. Quoted in Armin Hermann, *The Genesis of the Quantum Theory (1899–1913)* (MIT Press, 1971), p. 23.

15. $E = hv$, where E is energy, h is Planck's constant and v is frequency.

16. Flann O'Brien, *The Third Policeman* (Paladin, 1988), p. 69.

17. Quoted in Tony Hey and Patrick Walters, *The New Quantum Universe* (Cambridge University Press, 2005), p. 81.

18. www.almaden.ibm.com/vis/stm/corral.html. Image by M.F. Crommie, C.P. Lutz and D.M. Eigler, IBM Research Division, Almaden Research Center, California.

19. H.G.J. Moseley, 'The high frequency spectra of the elements, Part II', *Philosophical Magazine*, vol. 27, p. 703 (1914).

20. A neutral or uncharged atom contains the same number of electrons as there are protons in the nucleus. Most neutral atoms are very reactive as they have a very strong tendency to exchange electrons with other atoms. Negatively charged electrons are attracted and bound to the positively charged nucleus. The electric charge of a proton is equal but opposite to the electric charge on an electron. An electrically neutral atom will therefore contain the same number of electrons as there are protons in its nucleus.

21. In Moseley's apparatus, electrons were fired at a sample of the element under investigation. This bombardment knocked electrons out of the innermost shell of the atoms in the sample. The vacant slot in the shell would then be filled by another electron, with the emission of an X-ray photon. The X-rays produced in this way were then diffracted through a crystal and onto a photographic plate. The wavelength of the X-rays could be deduced from the angle through which they were diffracted, and this determined the energy of the photon and therefore the energy level of the innermost electrons of the atoms in the sample. The Bohr model of the atom was then used to deduce the charge on the nucleus that would match the measured energy level.

22. 'Le travail de Moseley substituait à la classification un peu romantique de Mendeléev une précision tout scientifique.' Quoted in Pais, *Inward Bound*, p. 229.

23. Richard Reeves, *A Force of Nature: The Frontier Genius of Ernest Rutherford* (Atlas Books, 2008), p. 91.

24. Quoted in Reeves, *A Force of Nature*, p. 95.

25. www.numericana.com/arms/rutherford.htm.

26. The most important feature of an electron's spin is completely inexplicable in terms of our intuitive understanding of the everyday world. If an electron's wave is rotated through a full 360°, the wave becomes minus its original value. A double rotation of 720° is required to return the electron wave to its starting value. For this reason electrons are assigned a spin of ½. This is an intrinsic property of the electron that cannot be altered.

27. There is such a thing as a square drum: the adufe is a square or rectangular tambourine-like instrument traditionally played by Portuguese women. For an illustration of this instrument, see www.nunocristo.com/adufe.html. David J. Benson, *Music: A Mathematical Offering* (Cambridge University Press, 2007), p. 117.

28. Only the modes of vibration that have the same cross section horizontally and vertically will not be paired up with a second wave with the same energy.

29. If the Schrödinger equation is solved for an electron in the hydrogen atom, the energy levels are found to stack up in the following sequence: 1, 4, 9, 16, … . Taking into account the two possible spin states of an electron, this means that the shells can be occupied by 2, 8, 18, 32, 50, … electrons. The hydrogen atom itself contains only a single electron, which will sink down into the first shell. In more complicated atoms, with several electrons, the electrons will repel one another. This increases the energy of the electron waves, which turns out to affect the waves in some of the shells more than it does the waves in others. The result is that the numbers of electrons in successive shells in multi-electron atoms form the following sequence: 2, 8, 8, 18, 18, 32, 32, … .

30. Richard Feynman, *QED: The Strange Theory of Light and Matter* (Penguin, 1990), p. 9.

31. Richard P. Feynman, Robert B. Leighton and Matthew Sands, *The Feynman Lectures on Physics*, vol. 1 (Addison-Wesley, 1963), Section 37–1.

4 QED

1. Goethe, *Faust*, part I, lines 3956–3959. www.poetryintranslation.com/ PITBR/German/Fausthome.htm.

2. This is the same effect that Faraday demonstrated with his homopolar motor, in which a wire carrying a flow of charge rotates around a magnet.

3. The techniques developed in Cambridge were later refined by Americans Robert Millikan and Harvey Fletcher, allowing them to make much more precise measurements of the charge on the electron.

4. T. Kuhn, Interview with Dirac, 1 April, 1962, American Institute of Physics, New York. Quoted in Abraham Pais, *Inward Bound: Of Matter and Forces in the Physical World* (Oxford University Press, 1988), p. 286.

5. Quoted in Pais, *Inward Bound*, p. 288.

6. 'Roundy', *Wisconsin State Journal*, 'P.A.M. issue' (April 31st [sic] 1929). Quoted in Silvan S. Schweber, *QED and the Men Who Made It: Dyson, Feynman, Schwinger and Tomonaga* (Princeton University Press, 1994), p. 18.

7. Hermann Weyl (1885–1955) was a leading German mathematician. He made important contributions to many areas of theoretical physics, including the formulation of general relativity and the application of group theory to quantum physics.

8. It is impossible to accelerate a body that possesses mass to the speed of light. Only massless particles, such as the photons of electromagnetic radiation, can travel at the speed of light.

9. Of course, Maxwell's equations could not work correctly if they were not relativistic, as they necessarily describe electromagnetic influences travelling at the speed of light.

10. Schweber, *QED and the Men Who Made It*, p. 70.

11. Anthony Zee, *Quantum Field Theory in a Nutshell* (Princeton University Press, 2003), p. 89.

12. Pais, *Inward Bound*, p. 403. Hess and his colleagues took a gold-leaf electroscope with them in their balloon to detect the radiation.

13. Pais, *Inward Bound*, p. 351.

14. C.D. Anderson, 'The positive electron', *Physical Review*, vol. 43, p. 491 (1933).

15. 'Thousands of cold anti-atoms produced at CERN', CERN Press release, 18 September 2002, http://press.web.cern.ch/press/Press Releases/Releases2002/PR09.02Eantihydrogen.html.

16. Quoted in Schweber, *QED and the Men Who Made It*, p. 112.

17. Another way to look at this is to consider the enormous amount of energy that would be required to push two tiny, negatively charged hemispheres together. The negatively charged hemispheres would repel each other, and the smaller they were, the closer together they would have to be pushed to form a sphere, and the greater the energy required to force them together.

18. Tomonaga and his team in Japan developed techniques for performing calculations in QED working in awful conditions after Japan's defeat in the Second World War. This work was completely independent of the simultaneous work by Schwinger and Feynman. Tomonaga's techniques turned out to be very similar to Schwinger's.

19. This technique is known as renormalisation.

20. The correct unit is known as the Bohr magneton. It is equal to the electric charge of the electron multiplied by Planck's constant and divided by twice the mass of the electron.

21. This passage is based on biographical material that Schwinger submitted to the Nobel Foundation at the time of his Nobel Prize in Physics award, in 1965. See Schweber, *QED and the Men Who Made It*, p. 276.

22. These words, spoken by Feld at Schwinger's sixtieth-birthday celebrations at UCLA in 1978, are quoted in Schweber, *QED and the Men Who Made It*, p. 296.

23. A more logical name for it would be 'electromagnetic force strength' or 'electromagnetic force constant'. Like many terms in physics, the fine structure constant is so called for historical reasons. It is often the case that concepts are named when they first appear in an early,

partially understood theory, and by the time the concept has become well understood and absorbed into mainstream physics the name has stuck. Unfortunately, the everyday meaning of such terms can be a poor guide to their meaning in physics.

24. Pais, *Inward Bound*, p. 457.

25. Schweber, *QED and the Men Who Made It*, p. 318.

26. Schweber, *QED and the Men Who Made It*, p. 398.

27. Quoted in Schweber, *QED and the Men Who Made It*, p. 408.

28. Schweber, *QED and the Men Who Made It*, p. 434. Quoted from Feynman's interview with Charles Weiner in 1966.

29. A Feynman diagram that shows a single photon being exchanged is called a tree diagram. When this diagram is evaluated it is found to correspond to a force between the electrons that is an inverse square law, so this diagram is equivalent to the classical physics of Maxwell. Feynman diagrams that contain loops are quantum corrections to the classical result. When they are evaluated they give small deviations from an exact inverse square law. This means that at very small distances the electric force is not exactly an inverse square law force.

30. Quoted in Schweber, *QED and the Men Who Made It*, p. 354.

31. Bohr objected that because of the quantum nature of particles, their trajectories could not be represented in the concrete way that they appeared to be in Feynman's diagrams.

32. David Kaiser, *Drawing Theories Apart: The Dispersion of Feynman Diagrams in Postwar Physics* (University of Chicago Press, 2005), p. 56. According to Dyson, Feynman performed 'the most amazing piece of lightning calculation I have ever witnessed'.

33. Schweber, *QED and the Men Who Made It*, p. 505.

34. Schweber, *QED and the Men Who Made It*, p. 454.

35. Interview with Schweber, 13 November 1980, cited in Silvan S. Schweber, 'Feynman and the visualization of space-time processes', *Reviews of Modern Physics*, vol. 58, p. 449 (1986).

36. Freeman Dyson, 'This side idolatry', foreword to Richard P. Feynman, *The Pleasure of Finding Things Out* (Penguin, 1999), p. vii.

37. Each two-loop diagram is proportional to $(\alpha/\pi)^2$.

38. David Kaiser, *Drawing Theories Apart*, p. 210.

39. Richard Feynman, *QED: The Strange Theory of Light and Matter* (Penguin, 2007), p. 117.

40. Feynman, *QED*, p. 117.

41. Donald H. Perkins, *Introduction to High Energy Physics*, 2nd edn (Addison-Wesley, 1982), p. 322.

42. G. Gabrielse, D. Hanneke, T. Kinoshita, M. Nio and B. Odom, 'New

determination of the fine structure constant from the electron g value and QED', *Physical Review Letters*, vol. 97, 030802 (2006).

43. Perkins, *Introduction to High Energy Physics*, p. 324.

44. More precisely, $2 \times 1.001\,159\,652\,180\,85(76)$ – an error of 0.76 parts per trillion.

45. Brian Carl Odom, 'Fully quantum measurement of the electron magnetic moment', thesis, Harvard University (October 2004). Odom gives the following expression for g, the magnetic moment of the electron:

$$g = 2[1 + C_1(\alpha/\pi) + C_2(\alpha/\pi)^2 + C_3(\alpha/\pi)^3 + C_4(\alpha/\pi)^4 + \cdots + a]$$

The values of the constants C, derived from Feynman diagrams with increasing numbers of loops, are:

$C_1 = 0.5$, 1 diagram (Schwinger)
$C_2 = -0.328\,478\,965\,579\,\ldots$, 7 diagrams (exact analytic)
$C_3 = 1.181\,241\,456\,587\,\ldots$, 72 diagrams (exact analytic)
$C_4 = -1.728\,3(35)$, 891 diagrams (numerical evaluation – integrals of 373 diagrams have been verified by more than one independent formulation, as of 2004)
$C_5 = ?$, 12,672 diagrams – not yet completed

The final term in the equation, a, groups together contributions to the magnetic moment from Feynman diagrams that include muons and other particles. These diagrams do not necessarily have many loops, which is why they are not lumped together with the other diagrams, but their contributions will still be very small because the masses of these particles are much bigger than the mass of the electron.

5 WE ARE STARDUST!

1. The steam engines and dark satanic mills of the Victorian era were powered by coal, but it was clear that the energy released by burning coal would not be sufficient to keep a ball of flame the size of the Sun burning for any great length of time.

2. H.G. Wells, *The World Set Free* (The Book Tree, 2007), p. 12.

3. Brian Cathcart, *The Fly in the Cathedral* (Farrar, Straus & Giroux, 2005), p. 137.

4. One of the characteristic features of quantum mechanics is that there are a number of energy rungs available to an electron in an atom. And the nucleus of an atom has similar properties. Indeed, we can

understand why some nuclei are radioactive by comparing them to atoms in which an electron has been excited onto a higher rung of the energy ladder. If an electron finds itself on a high-energy rung and there is a lower-energy rung available, it will fall down to the lower rung and simultaneously emit a photon whose energy is equal to the difference in energy between the two rungs. Each element has a characteristic spectrum of photons that is emitted or absorbed as electrons move around between the energy rungs of its atoms. Just as for an electron in an atom, the energy levels of the nucleus are well defined. Similarly, an atomic nucleus will be unstable if there is a lower energy level available to it. Whenever radioactive nuclear decay occurs, the amount of energy released is characteristic of that particular nuclear process. But, of course, nuclear processes involve much more energy than chemical reactions.

5. Steven Weinberg, *The Discovery of Subatomic Particles* (W.H. Freeman, 1990), p. 157.

6. If the energy that was released were being shared between the electron and a gamma ray, then it would be natural to expect the maximum energy of the electrons to be close to the total energy released, because the energy of the electrons would be greatest on those occasions when the total energy was shared out such that the electron had almost all the energy and the photon took away only a very small proportion of the energy. The total heat produced would then equal the number of beta decays multiplied by the maximum energy of the electrons. However, this was not what was found. The total heat produced turned out to be equal to the number of decays multiplied by the average energy of the electrons. So it was indeed only the energy that was being released along with the electrons that was heating the sample.

7. Abraham Pais, *Inward Bound: Of Matter and Forces in the Physical World* (Oxford University Press, 1988), p. 315.

8. The particles emitted in beta decay are now actually defined to be antineutrinos, rather than neutrinos, but we need not worry about this distinction here.

9. We have taken a neutrino with energy 1 MeV to be a typical neutrino.

10. John Updike, 'Cosmic Gall', in *Telephone Poles and Other Poems* (Knopf, 1960).

11. Fred Reines, 'Search for the free neutrino', in *Weak Neutral Currents: The Discovery of the Electro-weak Force*, edited by David B. Cline (Addison-Wesley, 1997), p. 2–63.

12. Dennis Silverman, 'The first detection of the neutrino by Frederick Reines and Clyde Cowan', www.ps.uci.edu/physics/news/nuexpt.html.

13. Pais, *Inward Bound*, p. 569.

14. These protons are the nuclei of hydrogen atoms.

15. Fred Reines, 'Search for the free neutrino', p. 2–63.

16. John Gribbin, *Stardust: The Cosmic Recycling of Stars, Planets and People* (Penguin, 2000), p. 76.

17. The hypothetical nucleus formed from two protons and no neutrons is almost but not quite stable, which is why it is necessary for one of the protons to simultaneously transform into a neutron via the weak force. If the strong force were just slightly stronger, nuclear fusion reactions would take place much more quickly because helium-2 would be stable.

18. Gribbin, *Stardust*, p. 87.

19. Stanisław Lem, 'The Sixth Sally', in *The Cyberiad* (Harvest, 1985), p. 150.

20. James B. Kaler, *Stars* (Scientific American Library, 1992), p. 140. Stellar lifetime is proportional to mass to the power −2.5. Stars of about 1.3 solar masses or more convert hydrogen into helium via a sequence more complicated than the proton–proton chain described in the main text. It is known as the CNO cycle, as it involves various isotopes of carbon, nitrogen and oxygen.

21. Simon Mitton, *Fred Hoyle: A Life in Science* (Cambridge University Press, 2011).

22. E.M. Burbidge, G.R. Burbidge, W.A. Fowler and F. Hoyle, 'Synthesis of the elements in stars', *Reviews of Modern Physics*, vol. 29, p. 547 (1957).

23. James B. Kaler, *The Hundred Greatest Stars* (Copernicus, 2002), p. 33.

24. One of the great achievements of B^2FH was that it used recently acquired data on the rates of nuclear reactions measured in the laboratory to calculate the abundance of different isotopes in the universe. As we have seen, although both carbon-12 and carbon-13 are stable, 99 per cent of carbon atoms are formed from carbon-12 nuclei. B^2FH explained how the relative abundances of the elements and their isotopes could be understood as resulting from the nuclear processes in which they were generated within stars.

25. Although the nucleus of an iron atom is the densest of all atomic nuclei, this does not mean that iron is the densest of all materials. The density of a material is equal to its mass divided by its volume, and the volume is determined by the size of the atoms and by how closely they are packed together. The size of atoms depends on the size of the electron orbitals, and this does not bear any close relationship to the density of the nucleus. The densest elements are iridium, osmium and platinum.

26. Adapted from the text as quoted in Colin A. Ronan, *The Shorter*

Science & Civilisation in China, vol. 2 (Cambridge University Press, 1978), p. 207.

27. Aldebaran lies on the same line of sight as the Hyades as seen from the Earth, but it is not actually a member of the cluster, being less than half as distant.

28. It is the radiation from the pulsar that is illuminating the nebula.

29. Once nuclear densities have been reached, the exclusion principle comes into play. Just as electrons obey the exclusion principle, so do neutrons: no two neutrons can exist in the same quantum state. This prevents the neutrons from being squeezed even further together and makes the nuclear matter stiff and resistant to further compression.

30. The Sun's radius is about 7×10^8 m. A neutron star's radius is about 1.5×10^4 m, and it is about 1.5 times as massive as the Sun (call it twice as massive) and about 0.25×10^{15} times as dense as the Sun. The Sun is slightly denser than water, at 1.4 g/cm^3, or 1.4 g/ml). A teaspoon holds about 5 ml. A spoonful of neutron star therefore has a mass of $5 \times 1.4 \times 0.25 \times 10^{15}$ g, or 2×10^{12} kg – 2 billion tonnes.

31. Neutron stars are very interesting and complex environments in which the electromagnetic, weak, strong and gravitational forces are all very important. They could be unique in that all four forces play an important role in their physics.

32. Cosmic rays are mainly protons and other atomic nuclei that have been blasted into space with extremely high energies in supernovae explosions.

6 ZEN AND THE ART OF QUARK DYNAMICS

1. Zhang Yunqi, *Weiqi de faxian* ('Discovering weiqi'), Internal document of the Chinese Weiqi Institute, Beijing (1991), p. 2.

2. Quoted at www.kiseido.com.

3. http://warp.povusers.org/pics/GobanPic.jpg.

4. Robert Marshak, 'The multiplicity of particles', *Scientific American*, vol. 186, p. 22 (1953).

5. 'Discovery of the pion – 1947', *CERN Courier*, June 1997, reproduced at http://fafnir.phyast.pitt.edu/particles/pion.html.

6. Robert P. Crease and Charles C. Mann, *The Second Creation* (Quartet Books, 1997), p. 170.

7. Strange particles can decay only by the strong force, and only if there are less-massive strange particles they can decay into. Strangeness can then be conserved, as it must be in strong interactions. The K mesons are the lightest of the strange particles, so they cannot decay by the strong force.

8. Quoted by Nicholas Samios, 'Early baryon and meson spectroscopy culminating in the discovery of the omega-minus and charmed baryons', chapter 29 of *The Rise of the Standard Model: Particle Physics in the 1960s and 1970s*, edited by Lillian Hoddeson *et al.* (Cambridge University Press, 1997), p. 526.

9. Willis E. Lamb, Jr, 'Fine structure of the hydrogen atom', Nobel Lecture, 12 December 1955, reproduced at http://nobelprize.org/nobel_prizes/physics/laureates/1955/lamb-lecture.pdf.

10. Isospin symmetry is described by the Lie group SU(2). The combined symmetry of isospin and strangeness is described by the Lie group SU(3). The octet is one of the representations of SU(3). The SU(2) group of the isospin symmetry is a subgroup of the SU(3) group of the Eightfold Way. The 2 in SU(2) indicates that isospin symmetry is due to the mixing of two quantum fields. Similarly, the 3 in SU(3) relates to the mixing of three quantum fields.

11. M. Gell-Mann, 'The Eightfold Way: A theory of strong interaction symmetry', US Department of Energy Report TID-12608 (1961); reprinted in Murray Gell-Mann and Yuval Ne'eman, *The Eightfold Way* (Perseus, 2000), p. 8.

12. Abraham Pais, *Inward Bound: Of Matter and Forces in the Physical World* (Oxford University Press, 1988), p. 514.

13. For instance, the xi-star and sigma-star resonances can decay to the xi and sigma particles without losing any strangeness. These decays are by the strong interaction and occur in around 10^{-23} seconds.

14. M. Gell-Mann, 'Strange particle physics. Strong interaction', in *Proceedings of the International Conference on High Energy Physics* (CERN, 1962), p. 805; reprinted in Murray Gell-Mann and Yuval Ne'eman, *The Eightfold Way* (Perseus, 2000), p. 87.

15. W.-M. Yao *et al.* [Particle Data Group], 'Review of particle physics', *Journal of Physics G: Nuclear and Particle Physics*, vol. 33 (2006). See the baryon summary tables at http://pdg.lbl.gov/2011/tables/rpp2011-tab-baryons-Omega.pdf.

16. Samios, 'Early baryon and meson spectroscopy ... ', p. 530.

17. 'Alternating Gradient Synchrotron', www.bnl.gov/bnlweb/history/AGS_history.asp.

18. James Joyce, *Finnegans Wake* (Faber, 1975), p. 383.

19. The atomic number is the number of protons in the nucleus of an atom. This number determines the positive charge on the nucleus, which in turn determines the number of electrons that must orbit the nucleus in order for the overall charge of the atom to be zero. And it is the arrangement of these electrons in an atom that determines its chemical properties. So the structure of the periodic table is

determined by the limited number of ways in which atoms can be built out of more elementary particles, namely protons and electrons. And the regularity in the composition of different atoms leads to the patterns in the periodic table.

20. George Johnson, *Strange Beauty: Murray Gell-Mann and the Revolution in Twentieth-Century Physics* (Vintage, 1999), p. 225.

21. Johnson, *Strange Beauty*, p. 43.

22. Some of the particles in the decimet have the same quark content as particles in the baryon octet. For instance, the delta-plus has the same quark content as the proton. This is because the quarks in the decimet particles, such as the delta-plus, all have their spin aligned in the same direction, giving these particles a spin of $3/2$, whereas the quarks in the proton and the other particles of the octet are not all aligned in the same direction, giving these particles a spin of $1/2$.

23. The quark composition of the particles can also be used to explain the extended lifetimes of particles such as the omega-minus. The weak force was first observed in the beta decay of unstable atomic nuclei. In beta decay a neutron is converted into a proton, and an electron is emitted. From Gell-Mann's quark model the neutron has quark content (d, d, u). The neutron is transformed into a proton, with quark content (d, u, u), when it undergoes beta decay, so one of the down quarks in the neutron must have been transformed into an up quark to form the proton. This suggests that quark flavour can change in weak interactions. If we postulate that the flavour of quarks remains unaltered in strong interactions, but that the weak interaction mixes the quark flavours, we can use the quark hypothesis to explain the longevity of the omega-minus. The omega-minus is composed of three strange quarks. The only way in which it can decay is for one of the strange quarks to transform into an up quark or a down quark. This can happen only in weak interactions, and this explains the longevity of the omega-minus.

24. Andrew Watson, *The Quantum Quark* (Cambridge University Press, 2004), p. 229.

25. Telephone conversation quoted in Crease and Mann, *The Second Creation*, p. 283.

26. Yves Klein, "'Ma position dans le combat entre la ligne et la couleur' (Paris 1958)'. Quoted in *Colour After Klein: Re-thinking Colour in Modern and Contemporary Art*, edited by Jane Alison (Black Dog Publishing, 2004), p. 36.

27. When an electron and a positron collide and annihilate, the energy released is available for the production of new particles. These particles may be quarks and antiquarks, or any other collection of particles whose creation does not violate the conservation of energy, electric

charge and all the other charges that are conserved in the interaction. This does not mean that the electron or the positron are composed of quarks.

28. W.N. Cottingham and D.A. Greenwood, *An Introduction to the Standard Model of Particle Physics* (Cambridge University Press, 2007), p. 13.

29. The exclusion principle applies to all matter particles, including electrons, protons, neutrons and quarks. This was first explained theoretically by Wolfgang Pauli in 1940, who demonstrated that it followed from the principles of quantum mechanics when combined with relativity. Pauli had originally postulated the exclusion principle seventeen years earlier, in 1923.

30. From Positron-Elektron-Tandem-Ring-Anlage, 'positron–electron tandem ring facility'.

31. About 10 per cent of the events include a third jet formed by a gluon. In about 1 per cent of events there are four jets, two of which are formed by gluons.

32. R.K. Ellis, W.J. Stirling and B.R. Webber, *QCD and Collider Physics* (Cambridge University Press, 1996), p. 70.

33. http://cdsweb.cern.ch/record/39449.

34. Ellis *et al.*, *QCD and Collider Physics*, p. 49.

35. Cottingham and Greenwood, *An Introduction to the Standard Model*, p. 171.

36. http://nobelprize.org/nobel_prizes/physics/laureates/1969/.

37. Johnson, *Strange Beauty*, p. 265.

7 THE MYSTERY OF THE SECRET SYMMETRY

1. Quoted in I.J.R. Aitchison and A.J.G. Hey, *Gauge Theories in Particle Physics*, 2nd edn (Institute of Physics, 1989), p. 393.

2. In terms of group theory, the spherical chamber is symmetrical under the group consisting of all rotations around any axis in three-dimensional space. After the hole has been punctured in the sphere, the symmetry is broken to the much smaller group consisting of rotations around just one particular axis. This is the axis that runs through the centre of the hole and the point on the sphere diametrically opposite the hole.

3. John Locke, 'Of probability', chapter XV of *An Essay Concerning Human Understanding*, book IV (1690), section 5. http://enlightenment.supersaturated.com/johnlocke/BOOKIVChapterXV.html.

4. Roald Z. Sagdeev, *The Making of a Soviet Scientist – My Adventures in*

Nuclear Fusion and Space from Stalin to Star Wars (Wiley, 1994), p. 43.

5. Walter Isaacson, *Einstein: His Life and Universe* (Pocket Books, 2008), p. 13.

6. John Emsley: *Nature's Building Blocks: An A–Z Guide to the Elements* (Oxford University Press, 2003), p. 270.

7. Emsley, *Nature's Building Blocks*, p. 155.

8. Sagdeev, *The Making of a Soviet Scientist*, p. 89.

9. Jean Matricon and Georges Waysand, *The Cold Wars* (Rutgers University Press, 2003), p. 99.

10. Metals are shiny because the free electrons can interact with light photons that impinge on the metal. If the electrons were bound to an individual atom they could interact only with a photon with exactly the right energy to promote the electron into a higher energy level in the atom.

11. 'The National Grid', Schoolscience, http://resources.schoolscience .co.uk/CDA/16plus/copelech5pg2.html.

12. The experiments in which superconductivity was discovered were performed by Gilles Holst under the direction of Kamerlingh Onnes. Holst was dissatisfied with the acknowledgment he received for his work and left the laboratory soon afterwards. He went on to become the director of the Philips Physics Laboratory.

13. Michael Tinkham, *Introduction to Superconductivity* (Dover, 2004), p. 2.

14. http://en.wikipedia.org/wiki/File:Magnet_4.jpg.

15. Its chemical formula is $La_{2-x}Ba_xCuO_4$; when $x = 0.15$ the critical temperature is 38 K.

16. James Gleick, 'Discoveries bring a "Woodstock" for physics', *New York Times*, 12 March 1987, reproduced at www.nytimes.com/1987/03/20/ nyregion/discoveries-bring-a-woodstock-for-physics.html.

17. 'Milestone superconductivity above 0 Celsius: The first material capable of superconductivity in permafrost', 26 December 2010, www.superconductors.org/276K.htm.

18. Marie Curie won two Nobel Prizes, but though the first was in physics, for her work on radioactivity, her second was in chemistry, awarded for her discovery of the elements radium and polonium.

19. The role of symmetry breaking in superconductivity was elucidated by Philip Anderson.

20. Some of the ideas that were devised by Higgs were also proposed by a number of other physicists at around the same time. They include Thomas Kibble, Gerald Guralnik, Carl Hagen, Robert Brout and François Englert. However, Higgs was the first to recognise that the symmetry-breaking mechanism required the existence of a new particle, now known as the Higgs particle.

21. Silvan Schweber, 'A historical perspective on the rise of the Standard Model', in *The Rise of the Standard Model: Particle Physics in the 1960s and 1970s*, edited by Lillian Hoddeson *et al.* (Cambridge University Press, 1997), p. 645.

22. As the Higgs force is part of the structure of the electroweak theory, it is not usually thought of as an additional force that must be unified with the electromagnetic, weak and strong forces. However, the exchange of the Higgs particle will produce a force in exactly the same way that photon exchange produces the electromagnetic force. But it will be an incredibly feeble force, much weaker than the weak force, simply because the Higgs particle is so massive.

23. 'So perhaps we humans should suspect that we live inside a superconductor ... Thus we come to suspect that the entity we call empty space is an exotic kind of superconductor.' Frank Wilczek, *The Lightness of Being: Mass, Ether, and the Unification of Forces* (Basic Books, 2008), p. 96.

24. Effectively, the Higgs field has two parts: one part is the constant background field, and the other part is variable. It is the excitations of the variable part that are the Higgs particles. All the particles that interact with the Higgs particles will also interact with the background Higgs field, and it is through these interactions with the background field that they gain their mass.

8 THE GRAND SYNTHESIS

1. Robert Rathbun Wilson, *Starting Fermilab*, Fermilab History and Archives Project, http://history.fnal.gov/GoldenBooks/gb_wilson2.html.

2. Quoted in Lillian Hoddeson, Adrienne W. Kolb and Catherine Westfall, *Fermilab: Physics, the Frontier and Megascience* (University of Chicago Press, 2008), p. 65.

3. Wilson, *Starting Fermilab*.

4. Wilson, *Starting Fermilab*.

5. Hoddeson *et al.*, *Fermilab*, p. 152.

6. Abraham Pais, 'Twenty years of physics: particles', *Physics Today*, vol. 21(5), p. 24 (May 1968).

7. Steven Weinberg, 'A model of leptons', *Physics Review Letters*, vol. 19, p. 1264 (1967).

8. 'And then the most unbelievable thing happened: nobody seemed to be aware of Weinberg's 1967 article, including Weinberg himself, who showed great interest in our work.' John Iliopoulos, 'What a

fourth quark can do', in *The Rise of the Standard Model: Particle Physics in the 1960s and 1970s*, edited by Lillian Hoddeson *et al.* (Cambridge University Press, 1997), p. 447.

9. Quoted in Peter Galison, 'The discovery of weak neutral currents', in *Weak Neutral Currents: The Discovery of the Electro-weak Force*, edited by David B. Cline (Addison-Wesley, 1997), p. 5.

10. Steven Weinberg, *Dreams of a Final Theory: The Scientist's Search for the Ultimate Laws of Nature* (Vantage, 1993), p. 96.

11. In the words of Sheldon Glashow, 'Tapestries are made by many artisans working together. The contributions of separate workers cannot be discerned in the completed work, and the loose and false threads have been covered over. So it is in our picture of particle physics.' From 'Towards a unified theory – threads in a tapestry', in *Nobel Lectures, Physics 1971–1980*, edited by Stig Lundqvist (World Scientific Publishing, 1992), p. 494. Also available at www.nobelprize.org/nobel_prizes/physics/laureates/1979/glashow-lecture.html.

12. Strictly speaking, this is a muon antineutrino.

13. Strictly speaking, this is an electron antineutrino.

14. François Rabelais, *Five Books of the Lives, Heroic Deeds and Sayings of Gargantua and His Son Pantagruel*, translated by Sir Thomas Urquhart of Cromarty and Peter Antony Motteux, book 1.iii (Penguin, 1955).

15. Helge Kragh, *Quantum Generations: A History of Physics in the Twentieth Century* (Princeton University Press, 2002), p. 256.

16. Beams of muon neutrinos were produced as the decay products of beams of pions.

17. John Iliopoulos, 'What a fourth quark can do'.

18. 'Charm' was a name that Glashow had originally used in a paper several years earlier.

19. All the particles in the second generation are unstable. They all decay via the weak force into the particles of the first generation. The first generation of particles are stable because there are no equivalent lighter particles for them to decay into. These are the particles that condense to form atoms, and the particles from which ordinary stable matter is composed.

20. SPEAR was the Stanford Positron Electron Asymmetric Rings; SLAC comes from the laboratory's original name, Stanford Linear Accelerator Center.

21. Martin Perl, 'The discovery of the tau lepton', in *The Rise of the Standard Model*, p. 79.

22. The bottomonium and charmonium mesons are keenly studied because they provide relatively simple systems for studying the colour force between the quark and the antiquark. Because bottom and

antibottom are so heavy, their motion is slow (compared with the speed of light), and this simplifies the analysis. It makes the mathematics of bottomonium very similar to the quantum mechanical analysis of the hydrogen atom that dates all the way back to Schrödinger's groundbreaking papers of 1926.

23. Weinberg, *Dreams of a Final Theory*, p. 3.
24. The antiproton accumulator is the machine that was added to the SPS to collect the antiprotons, which were then focused into a tight beam using stochastic cooling. This was essential for producing significant numbers of proton–antiproton collisions when the SPS operated in collider mode. For more information see http://cerncourier.com/cws/article/cern/28849.
25. www.nobelprize.org/nobel_prizes/physics/laureates/1984/.
26. R.R. Wilson's Congressional Testimony, April 1969, Fermilab History and Archives Project, http://history.fnal.gov/testimony.html.
27. Al Silverman, 'The magician: Robert Rathbun Wilson 1914–2000', *CERN Courier*, 7 March 2000, http://cerncourier.com/cws/article/cern/28180.
28. 'Physicists find first direct evidence for tau neutrino at Fermilab', Fermilab press release, 20 July 2000, www.fnal.gov/pub/presspass/press_releases/donut.html.
29. However, the existence of three generations offers the possibility that the symmetry between matter and antimatter might be violated in some reactions, and this might be related to processes that took place in the very early moments of the universe which resulted in the imbalance between the matter and antimatter in the universe, i.e. the fact that the universe is composed solely of matter. Without three generations, perhaps there would be no matter in the universe at all. The equal amounts of matter and antimatter produced in the early universe would have mutually annihilated, and all that would be left would be a radiation bath.
30. 'The Superconducting Super Collider project: A summary', http://web.archive.org/web/19990902043623/http://www.hep.net/documents/drell/apendixa.html (Appendix A of the Drell Panel Executive Report, US Department of Energy, High Energy Physics Advisory Panel's Subpanel on Vision for the Future of High Energy Physics, chaired by Sidney D. Drell, May 2004).
31. Leon Lederman with Dick Teresi, *The God Particle: If the Universe Is the Answer, What Is the Question?* (Delta, 1994).

9 THE ULTIMATE PRIZE

1. Quoted in Lilian Hoddeson, Adrienne W. Kolb and Catherine Westfall, *Fermilab: Physics, the Frontier and Megascience* (University of Chicago Press, 2008), p. 96.

2. www.cern.ch.

3. James Gillies and Robert Cailliau, *How the Web Was Born* (Oxford University Press, 2000), p. 221.

4. J.R.R. Tolkien, *The Fellowship of the Ring* (HarperCollins, 2007), p. 66.

5. The main sources for the data on the LHC that appear in this chapter are *CERN faq: The LHC Guide*, http://cdsweb.cern.ch/record/1092437/ files/; and Maria Spiropulu and Steinar Stapnes, 'LHC's ATLAS and CMS detectors', in *Perspectives on LHC Physics*, edited by Gordon Kane and Aaron Pierce (World Scientific Publishing, 2008), p. 25 (available at http://213.55.83.52/ebooks/physics/20591.pdf).

6. Ovid, 'Atlas transformed into a mountain', *Metamorphoses*, book IV. From the 1609 translation by John Dryden and others.

7. Information about ATLAS including video clips and simulated collision events is available on the CERN website at http://atlas.ch/.

8. *CERN faq: The LHC Guide*, p. 39.

9. All types of communication are limited by the speed of light, so no conceivable electronics within the detector would have sufficient time to read out the results of a collision event before the next event came in.

10. 'Hunt for the Higgs kicking into high gear', *Symmetry*, Fermilab/ SLAC, 16 February 2009 (2:40 am), www.symmetrymagazine.org/ breaking/2009/02/16/.

11. Karl Jakobs and Marcus Schumacher, 'Prospects for Higgs boson searches at the LHC', chapter 11 of *Perspectives on LHC Physics*, ed. Kane and Pierce.

12. According to the Standard Model, the Higgs mechanism is responsible for giving mass to the W and Z particles and the three generations of matter particles – the quarks and leptons. However, only part of the mass of a proton or neutron is due to the mass of its quark constituents. Most of their mass derives from the strong force holding the quarks together. This means that most of the mass of an atom is due to the strong force and not the Higgs mechanism.

13. Karl Jakobs and Marcus Schumacher, 'Prospects for Higgs boson searches at the LHC', in *Perspectives on LHC Physics*, edited by Kane and Pierce, p. 182.

14. Richard Webb, 'Should we worry about what the LHC is not finding?', *New Scientist*, 25 July 2011, www.newscientist.com/article/dn20729 -should-we-worry-about-what-the-lhc-is-not-finding.html.

15. Francis H. Cook, *Hua-Yen Buddhism: The Jewel Net of Indra* (University of Pennsylvania Press, 1977), p. 2.

16. 'Worldwide LHC Computing Grid', http://lcg.web.cern.ch/LCG/.

17. 'CERN experiments observe particle consistent with long-sought Higgs boson', CERN press release, 4 July 2012, http://press.web.cern.ch/press/PressReleases/Releases2012/PR17.12E.html.
18. The Higgs should reveal whether electrons (and quarks) are composite particles – the Higgs interaction is proportional to the mass of the particle. It will be a test of the theory, whether the strength of the interactions between the particles and the Higgs are of the right size in order to generate the whole mass of the particle or whether some of the mass is due to other effects.
19. Plato, *Symposium*, 191a–191b, 191c–191d.
20. Plato, *Symposium*, 192e–193a.

PLATE SECTION

i.   The octahedron is a regular polyhedron with eight triangular faces. The cuboctahedron is a semi-regular polyhedron with eight triangular faces and six square faces.
ii. http://commons.wikimedia.org/wiki/Category:Brocken_spectres.
iii. www.noao.edu/image_gallery/html/im0565.html.
iv. http://supercomputing.fnal.gov/SC2006/images/Atlas_0511013_01.jpg.

Further Reading

This book has covered many topics in a limited amount of space. The following books are accessible and engaging for the general reader. The website www.higgsforce.co.uk has been constructed to provide further information about this book and follow future developments in the hunt for the Higgs particle.

John Barrow, *The Book of Universes: Exploring the Limits of the Cosmos* (Norton, 2011)
John Barrow has written many thought-provoking books about modern physics. *The Book of Universes* surveys current speculations about the origin and evolution of the universe.

Stephen Blundell, *Superconductivity: A Very Short Introduction* (Oxford University Press, 2009)
There are not many books on low-temperature physics accessible to the general reader. This is a very good recent book about superconductivity.

John Emsley, *Nature's Building Blocks: An A–Z Guide to the Elements* (Oxford University Press, 2001)
John Emsley's books provide accessible and fascinating descriptions of many aspects of chemistry. *Nature's Building Blocks* looks at each element in turn and provides concise information about its history, properties and applications.

Graham Farmelo, *The Strangest Man: The Life of Paul Dirac* (Faber & Faber, 2010)
The Strangest Man is a highly acclaimed and long overdue biography of Dirac, one of the great physicists of the twentieth century.

Richard P. Feynman, *QED: The Strange Theory of Light and Matter* (Penguin, 2007)
Feynman wrote lucidly about many subjects. This is his explanation of quantum electrodynamics, written for the general reader and without any mathematics, but still essential reading for all physicists.

Richard P. Feynman and Ralph Leighton *'Surely You're Joking, Mr Feynman!' Adventures of a Curious Character* (Unwin, 1986)
These tales tell of Feynman's adventures, in his own words. The book is a very amusing insight into the career and character of one of the most original minds of the twentieth century.

Martin Gardner, *The Ambidextrous Universe: Left, Right and the Fall of Parity* (Pelican, 1964)
Inevitably, in a book such as *Higgs Force* there is insufficient space to cover everything of relevance. The most important feature of the Standard Model that has been left out is the surprising fact that the weak force distinguishes between left and right. *The Ambidextrous Universe* offers an engaging account of the discovery of this property of the weak force by physicists in the 1950s.

Brian Greene, *The Elegant Universe: Superstrings, Hidden Dimensions and the Quest for the Ultimate Theory* (Jonathan Cape, 1999)
The Elegant Universe is a very readable description of current speculations about the physics beyond the Standard Model and the possibility of an ultimate unified theory.

John Gribbin, *Stardust: The Cosmic Recycling of Stars,*

Planets and People (Penguin, 2000)
This very readable book describes the remarkable connections between the synthesis of the chemical elements in the stars and the existence of life in the universe.

Tony Hey and Patrick Walters, *The New Quantum Universe* (Cambridge University Press, 2003)
The New Quantum Universe is an excellent survey of quantum mechanics, its applications and its impact on our lives.

Lillian Hoddeson, Adrienne W. Kolb and Catherine Westfall, *Fermilab: Physics, the Frontier and Megascience* (University of Chicago Press, 2008)
This is a fascinating book about the history of Fermilab from its foundation on the Illinois prairie to its role as world-leading particle physics laboratory. It includes a lot of wonderful material about Robert Wilson, Fermilab's first director and one of the most colourful physicists of recent times.

James B. Kaler, *The Hundred Greatest Stars* (Copernicus, 2002)
This excellent account of astrophysics is brought to life by the descriptions of one hundred of the most interesting stars in the sky, many of which can be seen without a telescope.

Arthur Koestler, *The Sleepwalkers: A History of Man's Changing Vision of the Universe* (Penguin, 1989)
The Sleepwalkers is a wonderful account of the early history of celestial mechanics. The core of the book tells the story of Kepler and his great discoveries.

Manjit Kumar, *Quantum: Einstein, Bohr and the Great*

Debate About the Nature of Reality (Icon, 2009)
This is a good recent history of the development of quantum mechanics and the debate about its interpretation.

Simon Mitton, *Fred Hoyle: A Life in Science* (Cambridge University Press, 2011)
Simon Mitton has written a very entertaining biography of one of the great scientific figures of the last century.

Richard Rhodes, *The Making of the Atomic Bomb* (Simon & Schuster, 1988)
This is a superb account of nuclear physics and the development of nuclear weapons in the Manhattan Project.

Simon Singh, *Fermat's Last Theorem: The Story of a Riddle That Confounded the World's Greatest Minds for 358 Years* (Fourth Estate, 2002)
Simon Singh's wonderful book about the puzzles of pure mathematics includes accounts of the lives of Abel and Galois and the development of group theory – the mathematical study of symmetry.

Credits

All illustrations were created by Nicholas Mee and are © Nicholas Mee, with the exception of those listed below.

Front cover and Plate 13: Mural on the building above the ATLAS detector at CERN by Josef Kristofoletti. Copyright CERN, Geneva.

Figure 25: 'Quantum Corral'. Copyright M.F. Crommie, C.P. Lutz and D.M. Eigler, IBM Research Division, Almaden Research Center, California, USA.

Figure 26: The coat of arms of first Baron Rutherford of Nelson. Copyright John Campbell, University of Canterbury, Christchurch, New Zealand.

Figure 31: Photograph of the trail of a positron through Anderson's cloud chamber. C.D. Anderson, 'The positive electron', *Physical Review*, vol. 43, p. 491 (1933).

Figure 49: The original photograph in which the omega-minus particle was discovered. V. E. Barnes *et al.*, 'Observation of a hyperon with strangeness number three', *Physical Review Letters*, vol. 12, p. 204 (1964).

Figure 58: Reconstruction of a two-jet event at LEP. Copyright CERN, Geneva.

Figure 73: Schematic layout of the LHC complex at CERN. Copyright CERN, Geneva.

Figure 74: The inside of the LHC tunnel during construction: a worker inside the LHC tunnel. Copyright CERN, Geneva.

Figure 75: Cut-away diagram of the ATLAS detector. Copyright CERN, Geneva.

Plate 7: The Crab Nebula and Pulsar. Copyright N.A. Sharp/ NOAO/AURA/NSF.

Plate 8: The hypergiant star Eta Carinae. Copyright Hubble

Space Telescope, NASA.
Plate 9: Peter Higgs. Copyright University of Edinburgh, Peter Tuffy.
Plate 10: Main Ring and Main Injector of Fermilab. Photograph by Reidar Hahn, Visual Media Services, Fermi National Accelerator Laboratory. Copyright Fermilab.
Plate 11: 'Moonrise over the Highrise'. Photograph by Reidar Hahn, Visual Media Services, Fermi National Accelerator Laboratory. Copyright Fermilab.
Plate 12: Photograph of the inside of ATLAS. Copyright CERN, Geneva.

The following song lyrics are reproduced with permission from the copyright owners as indicated:
'Lucy in the Sky with Diamonds' (Lennon/McCartney), © 1967 Sony/ATV Music Publishing.
'Woodstock' (Mitchell), © 1969 Alfred Music Publishing.
'Strangely Strange, But Oddly Normal' (Pawle), © 1969 BMG.
'Quark, Strangeness and Charm' (Calvert/Brock), © 1977 Rock Music Company Limited.

Acknowledgments

I would like to thank my teachers at Reddish Vale School, and especially my physics teacher John Anslow, who introduced me to some of the most wonderful ideas ever dreamt up. The most intellectually stimulating year of my life was the one I spent studying Part III Mathematics at Cambridge University. I would like to thank the lecturers who gave me such an important insight into the joys of general relativity, quantum field theory and the Standard Model: Hugh Osborn, Peter Goddard, Ron Horgan, Alan MacFarlane, John Taylor, Ian Drummond, Gary Gibbons and John Stewart. Thank you to Nick Manton, who as my PhD supervisor influenced how I saw the physical universe in many ways, especially in the aesthetics of applying elegant mathematical models to the world around us. I would also like to thank everyone whose work I have discussed and the authors of the many books I have consulted along the way.

Writing a book can be a long and lonely pursuit, which makes all the encouragement offered by friends and family particularly welcome. The most enjoyable part of writing this book has been the many wonderful discussions about life, the universe and everything that I have had during its emergence. Thank you to all my friends and colleagues who read drafts of the manuscript and gave valuable feedback: Claire Ellis, Julie Cogill, Chris Chritchley, David Ruaune, Phil Trwoga, Dave Blakeman, Ben Allanach and Tony Mann.

A special thanks to the friends who offered assistance and encouragement throughout the entire course of the book's development. Thank you to Nigel Russell-Turner, Barry Phipps, Juliet Smith, Jane McAdam Freud and Karen Day for many stimulating conversations and unceasing moral support. Especial thanks to Jonathan Evans, who carefully read the whole manuscript and offered many valuable suggestions.

A warm thank you to Simon Singh for his careful reading of an early draft of part of the manuscript, but even more for his inspirational contributions to the art of science writing: no one has done it better. Simon's books, especially *Fermat's Last Theorem* and *The Code Book*, were the models I had in mind when attempting to write a captivating book about science.

Someone whose influence lies behind many of the artistic strands that always seem to appear in my work is John Eastwood. I would like to thank John for lots of stimulating discussions over the years and for his help during the design and construction of my websites.

Thank you to Kathryn Rudy, whose assistance was invaluable in polishing the manuscript for publication.

Enormous thanks to Debra Nightingale, whose perceptive comments influenced many aspects of this book from its inception.

Thank you to Dave Blakeman, Caroline Sanderson, Juliet Smith and everyone at Quantum Wave Publishing for working with me to produce this edition of the book. Thank you to Sally Coleman for her cover design, and thank you to John Woodruff for the rigour and efficiency of his copy-editing.

Thank you to Angie for her unceasing support and for putting up with me and my quarks.

And above all, thank you to my parents, who have always been incredibly supportive, well beyond the call of duty.

Index

Quantum Wave Publishing

Dear Reader,

Thanks for choosing *Higgs Force* – I hope you enjoyed it. We are a
new publishing company with the aim of giving you the best science
writing, presented in a clear and entertaining manner.

How Can We Help You?

To keep up to date with recent developments in science and our
products, simply visit our website *www.QuantumWavePublishing.com*
and join our mailing list. We'll email you occasional updates on
scientific developments, new publications and special offers.

Higgs Force is our first publication, but we plan a series of other books
and apps focused on popular science, and we'll send you details of
these as they become available.

How Can You Help Us?

We are always looking for innovative ways to improve the experience
for readers, and would welcome any suggestions to this end.
Please email us at *enquiries@QuantumWavePublishing.com* with your
ideas. You never know – your suggestion could well become reality!

If you have ever considered writing a popular science book,
or are interested in doing so, visit our authors' page at
www.QuantumWavePublishing.com/authors for submissions
guidelines and, if appropriate, to submit your proposal for
consideration. Without authors we would not have a business,
so we value your contribution.

And finally, if you are a bookshop owner, please let us know if there
is anything we could do to make your life easier. Simply email us at
enquiries@QuantumWavePublishing.com. We'd love to hear from you!

Once again, thanks for choosing this book!

Dave Blakeman
Managing Director
Quantum Wave Publishing

Nubble!

Nubble! is a PC game for testing and improving your children's arithmetic. It's deceptively simple but also requires strategy to win.

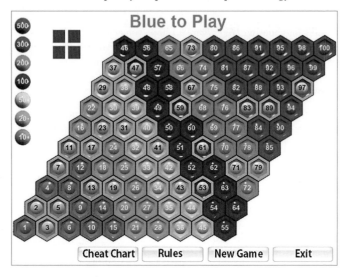

How It Works

Four dice are 'rolled', and the player has to combine the numbers using addition subtraction, multiplication and division to obtain a number between 1 and 100 that has not been chosen previously in the game. Some numbers are easy to find, while others are very tricky. The highest numbers score the most points.

What is a 'Nubble'?

A 'Nubble' is any group of three adjacent numbers that form a little triangle, and bonus points are scored for creating 'Nubbles'. If a 'Nubble' is created when a prime number is selected, even more bonus points are scored.

Games can be played against the computer or against one or more competitors.

How Do I Get It?

Nubble! Can be ordered online from our website (www.QuantumWavePublishing.com/Nubble), and can be downloaded so that you can start playing immediately. It works on all recent versions of Microsoft Windows.